Medicinal Plants
in Thailand

Volume II

Medicinal Plants
in Thailand

Volume II

Wongsatit Chuakul

Promjit Saralamp

Wichit Paonil

Rungravi Temsiririrkkul

Terry Clayton

DEPARTMENT OF PHARMACEUTICAL BOTANY
FACULTY OF PHARMACY, MAHIDOL UNIVERSITY

Department of Pharmaceutical Botany
Faculty of Pharmacy, Mahidol University

Medicinal Plants in Thailand
Volume II

Notice

Photographers

Santi Duangsub
Wongsatit Chuakul
Sompop Prathanturarug

Design

Siambooks and Publications Co., Ltd.
Tel. (662) 616-9215-8

First published in Bangkok in 1997
by Amarin Printing and Publishing Public Co., Ltd.

Tel. (662) 882-1010

ISBN 974-588-970-9

CONTENTS

PREFACE

Interest in medicinal plants among the general public in Thailand has grown drammatically during the 1980s through increased promotional support by both governmental and non-governmental organizations. In 1996, faculty members at the Department of Pharmaceutical Botany, Faculty of Pharmacy, Mahidol University published a book; Medicinal Plants in Thailand Volume I, an equivalent, English version of a book published four years earlier in Thai language. The book contained detailed information of 192 medicinal plants commonly used in Thailand. A lot of these medicinal plants are used as foods in Thai cooking and also as drugs for self medication in primary health care.

During the past few years we travelled extensively throughout the country conducting field survey, as well as interviewing traditional herbalists in order to collect more information about medicinal plants used in the rural areas. For these herbalists, medicinal plants play a far greater part in their everyday lives than most of us realize. The results of our research contained in this volume are information of 201 medicinal plants, completed with photographic illustrations. I certainly hope that more and more people will appreciate the immense richness of our culture in herbal medicine in Thailand and also will be able to benefit from the curative properties of these medicinal plants along with the correct plant utilization, which are the major objectives of this book.

Dr. Wongsatit Chuakul
Head of Department of Pharmaceutical Botany

Thai Folk Medicine

Folk healers in the Northeast use 'yaa fon' to treat constipation, diarrhea, fever, etc.

People of different races have different ways of dealing with their ailments, depending on their beliefs, natural environment, technological advancement and contact with other societies. These indigenous means of conquering physical illnesses gradually developed into systems of folk medicine or tribal medicine. It is not unusual for a society to possess more than one system of medicine. This is true for the Thai society, in which both the western medical system and indigenous Thai medical system are in common use. Only the Thai system will be considered here.

Thai Medical System

Based on geographical and cultural considerations, Thai medicine can be classified into two parts.

1. Thai Traditional Medicine

This indigenous system of medicine combines together the knowledge and practices involved in the diagnosis, prevention and treatment of physical and mental imbalances. It employs a systematic doctrine based on the Ayurvedic system of India. The knowledge and experience accumulated through generations of practice are well documented, usually in the form of old herbals and old writings. Part of the knowledge is naturally derived from folk medicine.

2. Thai Folk Medicine

This form of medicine is practised within the locality and the knowledge passed on within the community. As the result, it is usually characteristic of that particular area and is dependent upon the natural environment, and the social and cultural system of the tribe or community. In contrast to Thai traditional medicine, the knowledge derived at this level usually does not exist in a written form.

Before becoming a folk healer, the pupil must pay homage to his teacher by offering one set of 'bai-sri', 3 incent sticks, 5 pairs each of candles and flowers.

The Importance of Folk Medicine

While modern medicine has progressed a long way and the practice of modern medicine is accepted by a wide circle of the population where people are led to believe that modern medicine is synonymous with healthy living, a portion of the population still enlists the help of local herbalists for common ailments. From a survey conducted among health personnel in 1989, it was concluded that folk medicine still played an important role in primary health care where common ailments were involved. In effect, traditional healers should be of great help to modern doctors in alleviating their heavy burden. Therefore, folk medicine should be developed as an alternative medicine, particularly for people in rural areas.

Characteristics of Folk Medicine

In general, folk medicines are typified by the following characters :

1. Holistic Approach

Illnesses are regarded not only as physical abnormalities but also as abnormalities in the relationship between the people in the community as well as between the people and the environment. There is no dividing line between medicine, religion, and tribal laws. Religion and the medical system are inextricably interrelated and the health profile of the community reflects its way of life.

2. Diagnosis and Treatment of Diseases

Herbalists or folk medicine practitioners develop their diagnosis and treatment regimens by considering the interrelationship between human and human as well as human and environment. It could be said that folk medicine aims to cure social ills more than physical ills.

3. Harmony with the Tribal Way of Life

Folk medicine is in complete harmony with the way of life of the people. Since the villagers and the healers are from a similar background, having a similar educational and social status, there is no class difference. Moreover, their beliefs in the causes of illness are basically identical. Other advantages of folk medicine include the simplistic approach to the methodology and procedures used in the treatment of diseases. The most important factor, however, is that all members of the family can participate in the treatment.

Theoretical Approach to Folk Medicine

In accordance with the theory of Thai folk medicine, the causes of all illnesses may be divided into 2 categories.

1. Supernatural Causes

The supernatural causes of illnesses include evil spirits and deeds, evil forces and black magic. In addition to these, the positioning and regressions of certain stars and planets are believed to be the cause of some illnesses. The violation of tribal taboos may also result in physical or mental ills.

2. Naturalistic Causes

It is believed that the imbalance or loss of equilibrium of the four basic elements is the main cause of all ills. Contributing factors causing the imbalance are the age, natural and social environment of each individual.

Types of Folk Healers

Folk healers from different parts of the country may vary in accordance with the culture and beliefs of the people in that area. The word 'Moh' in Thai is used to denote both modern and traditional practitioners. To differentiate between the two systems, traditional healers are usually referred to as 'Moh Pan Boran' or 'Moh Ya Thai'. These healers may be classified into two main groups.

1. Folk Healers for Illnesses Caused by the Supernatural

Examples of this category of healers are

1.1 'Moh Song'

These healers will attempt to identify the causes of the illness by meditation. Once the causes are identified, the patients are treated accordingly.

1.2 'Moh Lum Pi Pha'

The healer in this category acts as a medium for the spirit of the sky to persuade the evil spirit to leave the body of the patient. This practice only exists in the northeastern parts of Thailand.

1.3 'Moh Su Kwan'

Kwan is defined as a vital part of one's conscious mind. If 'kwan' is missing, then illnesses may result. 'Moh Su Kwan' is responsible for the summoning of 'kwan'

Herbs are important raw materials and essential for the practice of folk medicine.

back into the body. A piece of white cotton thread or cloth is then tied to the patient's wrist after the ceremony has been performed by the healers.

2. Folk Healers for Illnesses Resulting from Natural Causes

2.1 'Herbalists'

'Moh Samun Phrai' or ' Moh Hark Mai', as they are called in the northeastern part of Thailand, make use of the knowledge inherited from their fore fathers. The principal remedies used are derived mainly from various kinds of medicinal plants. Occasionally, the herbalists will recite certain sacred passages as part of the healing process. Of all the different types of folk healers, herbalists play the most important role in the rural health care system.

2.2 'Bone Healers'

'Moh Kradook' or 'Moh Nam Mun' are those involved in the treatment of bone injuries. In cases of bone fractures, the remedial steps involve placing a bamboo cast over the affected area as well as the application of some ointment to assist healing.

2.3 Traditional Masseurs

'Moh Nuad' or 'Moh En' services patients with complaints involving muscle aches and pains. Skilled traditional masseurs are usually in great demand.

Treatment Procedure of Folk Healers

The steps involved in treatment by folk healers are as follows:

1. The setting up of a 'kai' or an offering consisting of five pairs of white flowers, five candles and a small amount of money.

2. The diagnosis. Different types of healers have their own ways of identifying their illnesses and their causes.

3. The treatment. After the healers have made the diagnosis, they will then prescribe a treatment

A complete 'kai' or offering

which may consist of the use of medicinal plants, the application of ointments, massage or the repelling of evil spirits, depending on their types of practice. If the treatment is ineffective, the patient will be advised to try different types of treatment.

4. The gratuity. After the patient has been cured, the kai or offering together with a type of cloth and a small amount of money or food or utility items will be given to the healer as a sign of gratitude.

The Survival of Folk Medicine in Thai Society

After western-style medicine gained popularity among the Thai people, folk medicine began to recede into the background, particularly in large cities where a modern life style has been adopted. However, many rural communities still adhere to the old way of life, including the use of folk medicine. It is not difficult to understand why these practices are still widespread in rural areas. Many factors contributed to the popularity of folk medicine and these are outlined below.

1. Its harmony with the way of life of the community. Folk healers normally employ a treatment procedure which fits in well with the way of life as well as social and economical status of the people. Emphasis is placed on the simplistic approach to treatment and participation of family members and relatives.

2. Types of illness. It is firmly believed that certain diseases can only be cured by folk healers, for instance, typhoid.

3. Causes of illness. Both the healers and the patients share a common belief that illnesses may result from either supernatural or natural causes.

4. Social structure. Strong family ties together with their respect for their elders contribute to the

choice of treatment people will resort to.

Other contributing factors include the following:

1. In accessibility to modern forms of treatment. In most cases, people will have to travel a long way to get to health centers. Therefore, they will tend to enlist the help of local healers.

2. Cheaper treatment costs.

3. Satisfaction with the service rendered.

4. Respectability. Most folk healers are well respected within the community.

5. The abundance of medicinal plants. As folk medicine relies chiefly on medicinal plants as therapeutic agents, continuous supply of these medicinal plants is vital to the survival of folk medicine.

However, folk medicine does not rely on any one factor for its survival, but a combination of these factors.

The Drawbacks of Folk Medicine

Although one of the strong points of folk medicine is its holistic approach to treatment, it also has its weaknesses, some of which are listed below.

1. The lack of systematic documentation. There is no record of the success or failure rates of the treatments prescribed by folk healers.

2. Satisfactory rating. The effectiveness of treatments cannot be measured by the satisfaction or expectation of the patients alone. Therefore, there is a need for concrete medical evidence to substantiate these claims.

Problems Facing Folk Medicine

The dwindling number of skilled folk healers pose a serious problem for folk medicine in recent times. There are no suitable successors to traditional doctors who have retired due to old age. Many of those who remain in the business lack the expertise of their predecessors. At the same time, there are quack doctors who make false claims in order to achieve financial gains. In addition, the adulteration of folk medicinal preparations with modern medicine is also a common practice. To make matters worse, the scarcity of certain medicinal plants in their natural habitats, which, without prompt intervention, will adversely affect the survival of folk medicine.

Future Trends in Folk Medicine

In the past, folk healers were responsible for the treatment of every type of ailment but their service has been superseded by modern medicine in many areas. Nowadays, these folk healers are limited to treating diseases which are incurable by modern medicine. This trend has, somewhat, reduced their roles in the health care system of the community. However, given the fact that these folk healers are in close contact with the local people and that folk medicine is in harmony with the peoples way of life, traditional practitioners should be considered as resource persons whose knowledge and skills need to be harnessed for the good of the people. Therefore, the strategy for the development of the national health system should be aimed at developing all systems of medicine concomitantly, thus giving the general public a free choice of the medical systems that are best suited to their needs.

THAI MEDICINAL PLANTS

Abelmoschus moschatus **Medic.**

MALVACEAE

Thai name ชะมดต้น, **Chamot Ton**

Shrub, 1-2 m high; branches hispid-hairy. Leaf simple, alternate, ovate or orbicular-ovate, 5-12 cm wide, 6-15 cm long; margin 3-5-lobed, chartaceous. Flowers in terminal or axillary cluster; corolla bright yellow with reddish purple base. Fruit capsule, oblong-ovoid; seeds musky.

Fresh leaf: topically apply for treatment of ringworm and tinea versicolor. *Flower*: treatment of involuntary and excessive discharge of semen during sleep. *Seed*: carminative.

Abutilon indicum Sweet

MALVACEAE

Chinese Bell Flower, Country Mallow

Thai name มะก่องข้าว, **Ma Kong Khaao**

Erect, branched shrub, 0.5-2 m high. Leaf simple, alternate, orbicular-ovate, broadly ovate or broadly cordate, 5-12 cm wide and long; both surfaces softly ashy-pubescent. Flower solitary, axillary, yellow. Fruit capsule, composed of 15-20 pubescent, shortly awned carpels.

Stem: blood tonic, carminative; promotes digestion and appetite. *Root*: element tonic, tonic; antipyretic; relieves cough, leukorrhea with unpleasant odor, gall bladder disfunction. *Leaf or whole plant*: decoction; diuretic, antidiabetic; treatment of urinary tract infection. *Leaf*: crush with small amount of water, apply to relieve abscess inflammation, toothache and gingivitis; boil with water for wet dressing. *Flower*: laxative.

Abutilon polyandrum G. Don

MALVACEAE

Thai name ก่องข้าวหลวง, **Kong Khaao Luang**

Erect, branched, pubescent shrub, 1-2.5 m high. Leaf simple, alternate, broadly ovate or cordate, 8-14 cm wide and long; both surfaces densely ashy-pubescent. Flower solitary, axillary, yellow, larger than *A. indicum*. Fruit capsule, composed of 12-18 pubescent carpels.

Stem: blood tonic, carminative; promotes digestion and appetite. *Root*: element tonic, tonic; antipyretic; relieves cough, leukorrhea with un-pleasant odor, gall bladder disfunction. *Leaf or whole plant*: decoction; diuretic, antidiabetic; treat-ment of urinary tract infection. *Leaf*: crush with small amount of water, apply to relieve abscess inflammation, toothache and gingivitis; boil with water for wet dressing. *Flower*: laxative.

Aeginetia indica Roxb.

BALANOPHORACEAE

Thai name ดอกดินแดง, **Dok Din Daeng**

Root parasite, leafless. Flower solitary; scape 15-40 cm long; corolla tube broad, incurved, purplish. Fruit capsule, sub-2-valved; seeds light yellow, very small.

Fresh or dried flower: crush with small amount of water and squeeze out the purple juice, used as food coloring agent.

Afgekia sericea Craib

FABACEAE

Thai name ถั่วแปบช้าง, Thua Paep Chaang

Climber, whitish pubescent. Leaves imparipinnate, alternate; leaflets 15-17, oblong or oblong-ovate, about 2 cm wide, 4 cm long. Inflorescence in terminal raceme; flowers pea-shaped, purplish pink; bracteoles pink. Pod flat, brown, with pale brown hairs.

Seed: fat tonic; strengthen tendons for slim and weak patient. *Root*: treatment of chickenpox.

Afzelia xylocarpa (Kurz) Craib

FABACEAE

Thai name มะค่าโมง, Makhaa Mong

Large tree, up to 30 m high. Leaves paripinnate, alternate; leaflets 3-5 pairs, oblong-ovoid, 4-5 cm wide, 5-9 cm long. Inflorescence in terminal panicle; calyx green, 4; corolla pink, 1. Pod flat, brown, woody-valved; seeds woody, brown, with orange aril.

Warty stem bark: anthelmintic; treatment of skin diseases and hemorroids. *Stem bark*: combine with equal amount of *Sindora maritima* stem bark, wrap with cloth, steam heat, apply to bruises or painful swelling; combine with equal parts of *Croton crassifolius* root for wound healing.

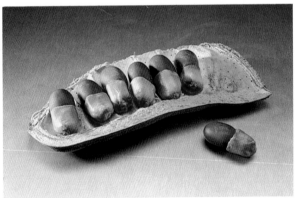

Aganonerion polymorphum Pierre ex Spire

APOCYNACEAE

Thai name ส้มลม, Som Lom

Lacticiferous climber. Leaves simple, opposite, ovate, lanceolate or elliptic, 2-4 cm wide, 4-6 cm long. Inflorescence in terminal, cymose panicle; flowers pink or reddish pink. Fruits a pair of follicles; seeds with whitish comas.

Root: decoction; relieves abdominal cramps and muscular pain.

Aganosma marginata G. Don

APOCYNACEAE

Thai name โมกเครือ, Mok Khruea

Lacticiferous liana, 5-10 m high. Leaves simple, opposite, oblong, 1.5-2.5 cm wide, 6.5-8.5 cm long. Inflorescence in terminal corymbiform cyme; flowers salver-shaped, white. Fruits a pair of follicles, linear, terete; seeds with caducous comas.

Stem: treatment of skin diseases and internal abscesses; decoction as antidiabetic. *Root*: tonic for patients recovering from fever; treatment of kidney and liver disfunctions; increasing menstrual discharge; decoction as laxative. *Leaf*: relieves muscular pain; externally apply to abscesses and hemorroids.

24

Allium tuberosum **Roxb.**

ALLIACEAE

Chinese Chive

***Thai name* กุยช่าย, Kui Chaai**

Annual herb; bulbs elongate, cylindric, 20-30 cm high. Leaf simple, basal, narrow linear, 2-4 mm wide, 15-30 cm long. Inflorescence in scapose umbel; perianth white; peduncle 30-45 cm long; bract membranaceous, spathaceous. Fruit 3-lobed capsule.

Seed: anthelmintic for threadworm and horse-whip worm; expels clotted menstrual discharge. *Stem and leaf*: grind with alcohol 28-40% and a little alum, filter, take for dysuria with urinary stones and gonorrhea. *Roasted seed*: grind to powder and mix with tang oil, soak with cotton, apply to dental carries; allow smoke of roasted seed to pass into the ear canal to kill small insects lodged there.

Altingia siamensis Craib

ALTINGIACEAE

Thai name ปรก, Prok

Large tree, 40-50 m high; bark rather smooth, brownish-grey, longitudinally peeling. Leaf simple, alternate, elliptic, oblong, ovate or ovate-lanceolate, 2-4 cm wide, 6-9 cm long. Inflorescence unisexual, monoecious; male inflorescence in racemose heads; female inflorescence in a single head. Fruit dry, subglobose, light brown.

Oleoresin and fragrant crystals from stem: flavoring agent for drugs and desserts.

Annona reticulata Linn.

ANNONACEAE

Custard Apple, Bullock's Heart

Thai name น้อยโหน่ง, Noinong

Small tree, 6-7 m high. Leaf simple, alternate, oblong or oblong-lanceolate, 4-6 cm wide, 12-15 cm long. Flowers solitary or 2-3-flowered fascicled, greenish yellow. Fruits aggregate, subglobose or heart-shaped, turned dark red when ripe.

Unripe fruit: antidiarrheal, antidysenteric, anthelmintic. *Fresh leaf*: crush with small amount of water and put on affected area to relieve painful swelling; treatment of tinea versicolor, ringworm, scabies, leprosy and yaws. *Unripe fruit and fresh leaf*: used as dye, black and blue color respectively. Leaves contain annoreticuin-9-one, squamone, solamin, annomonicin and rolliniastatin, which are toxic to cells and have potential for anticancer research.

Aporusa villosa (Lindl.) Baill.

EUPHORBIACEAE

Thai name เหมือดโลด, Mueat Lot

Tree, 8-10 m high; young parts yellowish-brown tomentose. Leaf simple, alternate, oblong, broadly oblong or elliptic-obovate, 6-10 cm wide, 10-16 cm long. Flowers apetalous, in axils of leaves or fallen leaves, unisexual, dioecious; male inflorescence in broadly bracteate clusters combined into groups of an elongate spike; female inflorescence in few-flowered, short, single spike. Fruit capsule, ovoid, irregularly dehiscent, turned reddish brown or orange when ripe.

Root or wood: grind with water, take as antipyretic.

Arfeuillea arborescens Pierre

SAPINDACEAE

Thai name คงคาเดือด, Khong Khaa Dueat

Tree, 8-20 m high. Leaves imparipinnate, alternate; leaflets 5-9, ovate or lanceolate, 2.5-4 cm wide, 4.5-7 cm long. Inflorescence in terminal panicle; flowers brownish. Fruit capsule, 3-winged.

.Wood: grind with water, take as anthelmintic. Stem bark: decoction as antipyretic; relieves internal fever and thirst; appetizer; boil with water, bathe to relieve ichting, neuralgia.

Argyreia nervosa (Burm. f.) Boj.

CONVOLVULACEAE

Thai name ใบระบาด, **Bai Rabaat**

Lacticiferous twining shrub, up to 10 m long, densely whitish or fulvous tomentose. Leaf simple, alternate, ovate or orbicular, 8-25 cm wide, 10-30 cm long, lower surface densely white or greyish tomentose. Flowers in axillary subcapitate cyme; corolla tubular to funnel-shaped, pink-purple, caducous, hairy midpetaline bands outside. Fruit berry, globose, yellowish brown, nearly dry.

Root: tonic, diuretic, aphrodisiac; treatment of allergic dermatitis with watery or pussy discharge, arthritis, obesity. *Leaf*: crush with water, apply to heal abscesses and wounds, treatment of skin diseases. *Fresh leaf juice*: used as eardrop to relieve inflammation.

Aristolochia pothieri Pierre ex Lec.

ARISTOLOCHIACEAE

Thai name กระเช้าถุงทอง Krachao Thung Thong

Climber, stem puberulous. Leaf simple, alternate, broadly ovate; margin entire or 3-lobed; lobes less than half the length of the lamina, 12-14 cm wide, 11-12 cm long. Inflorescence in axillary panicle; perianth reddish brown or brownish purple. Fruit septicidal capsule, ovoid.

Fresh root: slice into thin pieces and boil with water, take for longevity.

Asclepias curassavica Linn.

ASCLEPIADACEAE

False Ipecacuanha, Milk Weed, Bloodflower
Thai name ไฟเดือนห้า, Fai Duean Haa

Erect, perennial herb, 0.5-1.5 m high, lacticiferous; young stem and inflorescence minutely puberulous. Leaves simple, opposite, lanceolate, 2-3 cm wide, 6-13 cm long. Inflorescence in terminal, long-peduncled, umbel-like cyme; flowers red with yellow corona. Fruit follicle, fusiform; seeds brown with whitish coma.

Stem: element tonic, heart tonic; increases menstrual discharge; antiseptic for uterus infection after childbirth; treatment of fever with chills and unconciousness. *Leaf*: antileprotic, anthelmintic for roundworm; relieves abscess inflammation. *Root*: treatment of bruises. White stem latex inhibits growth of some fungi. Caution: contains asclepin which strongly stimulates the heart, beware of using internally.

Asparagus racemosus Willd.

ASPARAGACEAE

Thai name สามสิบ, Saamsip

Shrub, with subterranean rhizomes or tuberous roots and climbing stem; stems with recurved, very sharp spines, 1.5-4 m high. Leaf simple, alternate, reduced to narrowly linear scale, 0.5-1 mm wide, 10-36 mm long. Inflorescence in terminal or axillary raceme; perianth white. Fruit subglobose berry, red or reddish violet.

Root: tonic for pregnant women; liver and lung tonic. *Whole plant or root*: decoction; treatment of bloody vaginal discharge, treatment of goiter. Preserve in heavy syrup as dessert.

35

Averrhoa bilimbi Linn.

AVERRHOACEAE

Cucumber Tree, Bilimbi

Thai name ตะลิงปลิง, Taling Pling

Tree, 5-10 m high. Leaves imparipinnate, alternate; leaflets 21-45, ovate or oblong, 1.2-3 cm wide, 2-10 cm long, moderately densely pubescent beneath. Inflorescence in axillary, ramiflorous or cauliflorous panicle, pendulous, 5-20 cm long; flowers dark purple. Fruit berry, terete-obtusangular, greenish yellow, very sour.

Fruit: mucolytic; blood purifier. *Root*: relieves internal fever and thirst.

Baliospermum montanum (Willd.) Muell. Arg.

EUPHORBIACEAE

Thai name ตองแตก, **Tong Taek**

Small shrub, 1-2 m high; young shoots pubescent. Leaf simple, alternate; upper leaves lanceolate or elliptic, 3-4 cm wide, 6-7 cm long; lower leaves 3-5-lobed, elliptic-ovate, 7-8 cm wide, 15-18 cm long. Inflorescence unisexual or both sexes in the same inflorescence; male flowers numerous, at the upper part, apetalous, calyx greenish yellow, 4-5; female flowers at the lower part, apetalous. Fruit 3-valved capsule.

Leaf: decoction as purgative; soak in water, take as antiasthmatic. *Seed*: potent purgative. *Root*: boil or grind with water, take as purgative for hard fecal masses, itching with eczema, viscous and green sputum, mucous stool.

Bauhinia pulla Craib

FABACEAE

Thai name แสลงพันเถา, Salaengphan Thao

Tendrilled climber, up to 5 m high; young branches greyish pubescent. Leaf simple, alternate, ovate, bifid, up to 14 cm wide, 15 cm long; stipules falcate. Inflorescence in terminal or axillary raceme, up to 20 cm long; flowers greenish. Pod flat, woody-valved, dehiscent, velvety, strap-shaped.

Stem: decoction or powder, take as detoxifying agent. *Seed*: anthelmintic, diaphoretic, antipyretic; relieves internal fever; abscess healing.

Biophytum sensitivum DC.

OXALIDACEAE

Thai name กระทืบยอบ, **Krathuep Yop**

Erect, unbranched, annual herb; stem finely puberulous, dull red-brownish, up to 20 cm high. Leaves paripinnate, in a compact spiral at the tip of the stem, 6-11 cm long; leaflets 8-12 pairs, suborbicular, elliptic or obovate, 4.5-10 mm wide, 6-12 mm long. Inflorescence terminating the stem; peduncle 3-10 cm long; corolla tube light green, corolla lobes yellow with some red vertical steaks at the base inside. Fruit ellipsoid capsule, pale light greenish.

Whole plant: decoction; antipyretic, diuretic, antidiabetic; treatment of dysuria with bloody urine; crush with water, topically apply to abscesses, chronic wounds, scorpion bites. *Stem or leaf*: antihiccup; treatment of sorethroat; detoxifying agent. *Root*: treatment of gonorrhea and dysuria with urinary stones.

Blumea balsamifera (Linn.) DC.

ASTERACEAE

Camphor Tree

Thai name หนาดใหญ่, Naat Yai

Shrub, very aromatic, smelling of camphor, 1-4 m high. Leaf simple, alternate, oval-oblong; both surfaces densely tomentose, 2-20 cm wide, 8-40 cm long. Inflorescence in terminal or axillary head; corolla yellow. Fruit achene, slightly curved, angular, 5-10-ribbed, thinly shortly white-hairy.

Leaf: antiflatulent, carminative, diaphoretic, expectorant; relieves abdominal pain. *Slightly dried leaf*: smoke like a cigarratte as antiasthmatic; treatment of infectious rhinitis with pus. Leaves contain cryptomeridion which acts as smooth muscle relaxant and relieves bronchospasm.

Caesalpinia mimosoides Lamk.

FABACEAE

Thai name ช้าเลือด, Chalueat

Erect or climbing tree, densely hispid and bristly on all parts; stipules caducous. Leaves bipinnate, alternate, 25-40 cm long; pinnae 10-30 pairs; leaflets 10-20 pairs, about 4 mm wide, about 10 mm long. Inflorescence in terminal raceme; flowers yellow. Pod bladder-like.

Fresh young twig: taken as vegetable for fainting and hypotension.

Cajanus cajan Millsp.

FABACEAE

Angola Pea, Congo Pea, Pigeon Pea
Thai name ถั่วแระ Thua Rae

Erect shrub, 1-3.5 m high. Leaves pinnately trifoliolate, alternate, oblong-lanceolate; upper surface densely short-hairy; lower surface clothed with longer hairs, 1-3.5 cm wide, 1.5-10 cm long. Inflorescence in axillary peduncled raceme; flowers pea-shaped, yellow, often tinged with reddish brown. Pod linear, septate.

Root or seed: diuretic, antipyretic; treatment of nephritis with dark yellow or bloody urine; detoxifying agent. *Stem or leaf*: dispels gas; treatment of tendon disfunction. *Leaf*: antidiarrheal; relieves cough. *Leaf juice*: topically apply to aphthous ulcer or wounds in the ear. *Whole plant*: treatment of bloody vaginal discharge, fever during menstruation. Seed extract exhibits hypoglycemic activity in animal tests.

Calanthe cardioglossa Schltr.

ORCHIDACEAE

Thai name เอื้องน้ำต้น, Ueang Namton

Terrestrial orchid with sympodial pseudobulbs. Leaf simple, alternate, oblong-oblanceolate, 4-7 cm wide, 15-25 cm long. Inflorescence in axillary raceme; flowers pink or purplish red. Fruit capsule, linear-obovoid.

Bulb-like aerial stem: decoction; tonic.

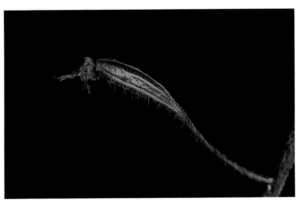

Calophyllum inophyllum Linn.

CLUSIACEAE

Alexandrian Laurel

Thai name กระทิง, Kra Thing

Medium-sized tree, 20-25 m high. Leaves simple, opposite, elliptic or obovate-oblong, 2.5-3.5 cm wide, 4-6 cm long. Inflorescence in upper axillary raceme, 4-6-flowered; corolla white; stamens yellow, numerous, fragrant. Fruit fleshy, globose, pulpy.

Flower: an ingredient in Ya-hom, formula for fainting; heart tonic. *Fixed oil from seed*: topically apply to relieve muscular pain, joint pain, sprains and swelling.

Calotropis gigantea (Linn.) R. Br. ex Ait.

ASCLEPIADACEAE

Crown Flower, Giant Indian Milkweed

Thai name รัก, Rak

Erect shrub, with white milky juice, 1-3 m high. Leaves simple, opposite, oblong or oblong-obovate, 4-15 cm wide, 8-30 cm long, both surfaces white tomentose. Inflorescence in axillary peduncled, many-flowered cyme; flowers lilac or white with corona. Fruit a pair of follicles; seeds with short-hairy comas.

White latex from stem: potent purgative, anthelmintic; topically apply to ringworm and tinea versicolor; relieves toothache, earache. *Stem bark*: emetic; treatment of allergic dermatitis with watery or pus discharge. *Flower*: promotes digestion and appetite; cough and asthma remedy. White latex contains heart stimulating substance, needs more research on toxicity.

45

Cananga latifolia Finet et Gagnep.

ANNONACEAE

Thai name สะแกแสง, Sakae Saeng

Tree, 10-20 m high; young twigs dark-colored tomentose. Leaf simple, alternate, broadly ovate-oblong; upper surface pubescent, 8.5-12 cm wide, 9-15 cm long. Flowers in axillary cluster; corolla light green, fragrant. Fruits aggregate, each oblong.

Wood: topically used for skin diseases, such as ringworm, tinea versicolor, leprosy, allergic dermatitis with watery and pussy discharge; scrape to scaly pieces, mix with tobacco, smoke as a cigarrette to treat infectious rhinitis with pus. *Leaf*: treatment of chronic wounds. *Wood, root or whole plant*: antipyretic.

Capparis micracantha DC.

CAPPARACEAE

Thai name ชิงชี่, Chingchee

Shrub or small tree, 1-6 m high; branches glabrous; thorns directed upwards. Leaf simple, alternate, oblong, elliptic or ovate, 4-10 cm wide, 8-24 cm long. Inflorescence axillary or supra-axillary; flowers white with yellow streak at first and turned brownish purple later. Fruit fleshy, ellipsoid, globose or elongated, turned red when ripe.

Root: carminative; treatment of chronic infected skin diseases. *Stem*: crush with small amount of water and topically apply to relieve sprains and swelling. *Leaf*: used for muscular cramps; boil with water, drink or bathe to relieve fever with chronic vesicular skin diseases; smoke to treat bronchitis. *Root or leaf*: antiasmatic; treatment of chest pain, fever with vesicular skin diseases, such as measles.

Cardiospermum halicacabum Linn.

SAPINDACEAE

Balloon Vine, Heart Pea

Thai name โคกกระออม, **Khok Kra Om**

Annual, climbing herb. Leaves ternately pinnate, alternate, 3-7 cm long; leaflets deltoid or ovate, 1-1.5 cm wide, 4-5 cm long. Inflorescence in axillary, long-stalked corymb or raceme with a pair of simple coiled tendrils near the base; flowers white. Fruit capsule, broadly pyriform, 3-cornered, membranous, 3-celled; seeds spherical, black with white heart-shaped accrescence.

Leaf: antiasthmatic, diuretic. *Leaf juice*: cough remedy. *Stem*: antipyretic. *Flower or leaf juice*: increases menstrual discharge. *Fruit*: crush with small amount of water and apply to relieve burns. *Whole plant*: antiasthmatic; treatment of arthritis. In animal tests, leaf extract possesses hypotensive and anti-inflammatory properties.

Careya sphaerica Roxb.

BARRINGTONIACEAE

Thai name กระโดน, **Kradon**

Medium-sized tree, deciduous, 8-20 m high. Leaf simple, alternate, obovate, 6-12 cm wide, 12-20 cm long. Flower solitary, terminal or above the leaf-scars; corolla light green with pinkish margins; stamens white, numerous, filament bases reddish pink and connate. Fruit drupaceous, globose.

Flower: tonic for post-labor. *Fruit*: promotes digestion. *Leaf*: wound healing. *Stem bark*: astringent for wounds; relieves sprains, fatigue; anti-inflammation from snake bite. Caution: not recommended for poisonous snake bite.

Carissa carandas Linn.

APOCYNACEAE

Carunda, Christ's Thorn

Thai name หนามแดง, **Naam Daeng**

Lacticiferous shrub, 2-3 m high; branches spiny. Leaves simple, opposite, obovate, elliptic, oblong, 1.5-4 cm wide, 3-7 cm long. Inflorescence in axillary peduncled cyme; corolla-lobes white with reddish-pink tube. Fruit berry, ovoid, turned dark purple when ripe.

Wood: fat tonic and strengthens tendons for slim patients; element tonic. *Fresh leaf*: decoction; antidiarrheal, antipyretic; relieves pain in the ear, mouth or throat. *Ripe and unripe fruits*: antidiarrheal, antiscurvy. *Fresh root*: decoction; anthelmintic, antidiarrheal, element tonic, appetizer; crush with 28-40% alcohol and apply to heal wounds or to relieve itching. Caution: Root contains cardiac glycoside, beware of using internally.

Carissa cochinchinensis Pierre

APOCYNACEAE

Thai name หนามพรม, **Naam Phrom**

Lacticiferous shrub, 4-5 m high; branches thorny. Leaves simple, opposite, obovate, 1.5-2.5 cm wide, 2.5-4 cm long. Flowers in terminal cymose cluster; corolla white, tubular, fragrant. Fruit berry, ellipsoid, turned dark purple when ripe.

Wood: tonic.

51

Casearia grewiaefolia Vent.

FLACOURTIACEAE

Thai name กรวยป่า, Kruai Paa

Shrub or tree, 2-20 m high. Leaf simple, alternate, oblong or ovate-oblong, 3-6 cm wide, 8-10 cm long. Flowers clustered in axillary dense-flowered fascicles, white to yellowish green. Fruit succulent capsule, compressed ellipsoid, yellow when ripe; seed with orange-red aril.

Root: antidiarrheal; treatment of liver disfunction. *Bark*: element tonic, tonic. *Flower*: antipyretic. *Leaf and flower*: detoxifying agent for infections. *Leaf*: treatment of skin diseases such as ringworm, tinea versicolor, scabies, etc.; smoke to treat nose infections, rhinitis; fry in vegetable oil, topically apply to relieve wounds and skin infections. *Seed*: relieves hemorrhoids; fish poisoning. *Fixed oil from seed*: treatment of skin diseases.

Cassia bakeriana Craib

FABACEAE

Wishing Tree, Pink Shower

Thai name กัลปพฤกษ์, **Kanpaphruek**

Tree, up to 10 m high; all younger parts densely hairy. Leaves paripinnate, alternate; leaflets 5-7 pairs, oblong-oblanceolate, 1.5-3 cm wide, 6-8 cm long; young leaflets light-brown velvety pubescent. Inflorescence in axillary raceme, 5-12 cm long; flowers pinkish. Pod terete, softly grey to brownish velvety pubescent.

Seed aril: mild laxative for children.

Cassia sophera Linn.

FABACEAE

Thai name ผักหวานบ้าน, **Phak Waan Baan**

Shrub, 1-3 m high. Leaves paripinnate, alternate; leaflets 4-9 pairs, lanceolate, 1-2 cm wide, 2-5 cm long; stipules ovate, caducous. Inflorescence in axillary, few-flowered corymb; flowers yellow. Pod swollen, nearly straight.

Root: relieves internal fever; an ingredient in Yaa-khieo, famous antipyretic formula; boil with water and bathe to relieve itching and to heal wounds.

Senna timoriensis (DC.) H.S. Irwin et R.C. Barneby

(*Cassia timoriensis* DC.)

FABACEAE

Thai name ขี้เหล็กเลือด, Kheelek Lueat

Small tree, up to 10 m high; young branches yellowish and golden hairy. Leaves paripinnate, alternate; stipules large; leaflets 10-20 pairs, oblong, 1-1.5 cm wide, 2-6 cm long. Inflorescence in axillary dense raceme, 10-30 cm long; bracts caducous; flowers yellow. Pod flat, dehiscent.

Wood: blood tonic, diuretic; ingredient in dismenorrheal formula; dispels stagnant blood; treatment of kidney disfunction, scrotal enlargement, waist or lower abdominal pain. *Stem bark*: treatment of scabies.

Cassytha filiformis Linn.

CASSYTHACEAE

Love Vine

Thai name สังวาลย์พระอินทร์, **Sangwaan Phra In**

Parasite; stem yellowish-green, filiform, twining, scrambling over other plants; young parts yellow tomentose. Leaf none or of minute scales. Inflorescence in axillary spike; flowers globose, yellowish-green. Fruit fleshy, obovoid or globose, white at maturity.

Whole plant: liver and kidney tonic; treatment of common colds, cough with bloody sputum, dysuria with urinary stones, urinary tract infections, jaundice, dysentery, nosebleed, chronic wound inflammation, burns. For veterinary use, mix with lime water as bovine anthelmintic. In animal experiments, whole plant extract acts as diuretic, antipyretic, central nervous system depressive, hypotensive with respiratory depression. High doses cause convulsion and death.

Catunaregam tomentosa (Bl. ex DC.) Tirveng.

(*Randia dasycarpa* Bakh. f.)

RUBIACEAE

Thai name เคต, Khet

Small tree, 4-8 m high; branches long thorny. Leaves simple, opposite, broadly ovate or ovate-elliptic, 3-5 cm wide, 5-7 cm long, pale coloured and pubescent beneath; stipules interpetiolar. Flower solitary, axillary; corolla whitish, fragrant. Fruit dry, dehiscent, ovate.

Mature fruit: crush, agitate with water, use as shampoo.

Cayratia trifolia (Linn.) Domin.

VITACEAE

Thai name เถาคันขาว, **Thao Kan Khaao**

Climber, 2-20 m long, with tendrils. Leaves digitately trifoliolate, alternate; leaflets elliptic or lanceolate, 1-4 cm wide, 2-8 cm long. Inflorescence in axillary, long-peduncled, corymbiform cyme; flowers light green. Fruit berry, compressed-globose, turned black when ripe.

Leaf or root: antipyretic, astringent. *Stem*: expectorant, carminative, blood purifier; relieves vertigo, fainting, internal bruises; expels mucus via rectum. *Leaf*: antiscorbutic; crush with small amount of water and locally apply to nasal wounds; counter-irritant for sprains; heat and put on boils to relieve inflammations.

Celastrus paniculatus Willd.

CELASTRACEAE

Thai name กระทงลาย, **Kra Thong Laai**

Scandent shrub. Leaf simple, alternate, elliptic or elliptic-oblong, 2.5-4 cm wide, 6-8 cm long. Inflorescence in terminal or axillary panicle or raceme; flowers green. Fruit capsule, globose or ovoid; seeds with reddish brown aril.

Root: antimalarial. *Wood*: treatment of tuberculosis. *Fruit*: antiflatulent, blood tonic, antivenum (no scientific support); relieves fainting. *Leaf*: antidysenteric. *Seed*: crush with small amount of water and apply to relieve joint and muscular pains, paralysis. *Fixed oil from seed*: diaphoretic; treatment of beriberi. *Stem bark*: antidysenteric. *Stem*: decoction; diuretic; treatment of kidney disfunction.

Cerbera odollam Gaertn.

APOCYNACEAE

Thai name ตีนเป็ดทะเล, Teenpet Thale

Lacticiferous tree, 6-17 m high. Leaf simple, alternate, obovate-lanceolate or oblong-lanceolate, 4.5-7 cm wide, 12-31 cm long. Inflorescence in terminal, peduncled cyme; flowers white with yellowish center. Fruit drupelets, ellipsoid, turned dark purple when ripe.

Stem bark: laxative, antipyretic; treatment of dysuria with urinary stones. *Wood*: used for paralysis. *Flower*: relieves hemorrhoids. *Root*: expectorant. *Leaf*: externally used for ringworm and tinea versicolor. *Fixed oil from seed*: treatment of scabies; hair tonic. *Outer bark*: treatment of tinea versicolor. *Seed*: fish poisoning. Latex and seed contain heart toxic substances, cerberoside and thevobioside. Latex, leaf and fruit cause vomiting and diarrhea; high doses may cause death.

Chromolaena odorata (Linn.) King et Robins.

(*Eupatorium odoratum Linn.*)

ASTERACEAE

Thai name สาบเสือ, Saap Suea

Erect, branched, annual herb, up to 1.5 m high; younger parts densely pubescent. Leaves simple, opposite decussate, ovate, sparsely pubescent on both sides, 2-6.5 cm wide, 5.5-11.5 cm long. Inflorescence in terminal corymbose panicle; flowers white with lilac hue. Fruit achene, linear, flat, sparsely puberulous.

Fresh leaf: tropically apply to stop bleeding; contains 4,5,6,7-tetramethoxyflavone and calcium, which decrease blood clotting time. *Whole plant*: insecticide. *Root*: an ingredient in antimalarial formula. Chloroform and acetone extracts of stem and leaf inhibit the growth of *Staphylococcus aureus* and *Bacillus subtilis*, which cause pus formation.

Cibotium barometz J. Smith

DICKSONIACEAE

Thai name ละอองไฟฟ้า, **La Ong Faifaa**

Rhizomatous fern, very densely covered with golden yellow hairs, 1.5-2.5 m high. Leaves bipinnate, up to 2 m long; leaflets linear-lanceolate, deeply pinnatifid, 1.5-2.5 cm wide, 10-15 cm long. Sori terminal on lower veins, parallel to edge of lobes.

Golden colored hairs from rhizome: can be used fresh or dried and powdered, dirrectly apply to stop bleeding, especially from leeches and lacerations.

63

Cinchona succirubra Par.

RUBIACEAE

Thai name **ควินิน, Quinin**

Tree, 8-15 m high. Leaves simple, opposite, elliptic-lanceolate, 7-10 cm wide, 10-18 cm long; stipules lanceolate, interpetiolar. Inflorescence in terminal, short-peduncled, few-flowered cyme; corolla white or pink, tubular. Fruit capsule, linear-ellipsoid, turned reddish brown when ripe.

Stem bark: antimalarial. Stem and root bark contain many alkaloids. Quinine has been used as an antimalarial and quinidine for cardiac arrhythmia.

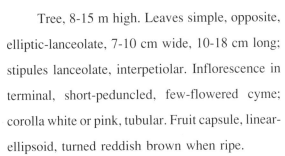 

Cladogynos orientalis Zipp. ex Span.

EUPHORBIACEAE

Thai name เจตพังคี, **Chettaphangkhee**

Shrub, white-stellate-hairy, about 2 m high. Leaf simple, alternate, ovate or elliptic, 3-8 cm wide, 6-15 cm long; lower surface densely white-pubescent and floccose-stellate-hairy. Inflorescence in axillary cluster, unisexual, monoecious, apetalous. Fruit schizocarp, 3-lobed, white-wooly.

Root: carminative, antiflatulent; relieves stomach pain; crush with lime water or mix with assa-foetida or camphor, apply to child's abdomen to relieve flatulence. *Whole plant*: decoction or powder or macerate in 28-40% alcohol as antiflatulent, antidiarrheal. *Root*: decoction; element tonic, heart tonic, appetizer.

Clausena harmandiana Pierre ex Guill.

RUTACEAE

Thai name ส่องฟ้าดง, Song Faa Dong

Undershrub, 20-50 cm high. Leaves imparipinnate, alternate; leaflets 4-7, ovate-oval-oblong, 2.5-4 cm wide, 4-8 cm long; leaf surface pellucid-punctate. Inflorescence in terminal panicle; flowers yellowish-white. Fruit berry, globose.

Root: decoction; antipyretic, antiflatulent; relieves headache, bronchitis; food poisoning.

Cleome viscosa Linn.

CLEOMACEAE

Thai name ผักเสี้ยนผี, **Phak Sian Phee**

Annual herb, up to 1 m high; densely glandular pubescent and viscid throughout. Leaves palmately compound, alternate; leaflets 3-5, obovate or ovate, 1-1.5 cm wide, 1.5-4.5 cm long. Inflorescence in terminal raceme; flowers yellow. Fruit silique, dehiscing from the base.

Whole plant: antiflatulent, antidiarrheal, antipyretic, fire element tonic; treatment of abdominal pain, abscess or infected inflammation, tumor and cancer in lung, intestine or liver. *Leaf*: urine disfunction. *Flower*: antiseptic. *Fruit*: anthelmintic. *Root*: treatment of uterine infection for post-labor.

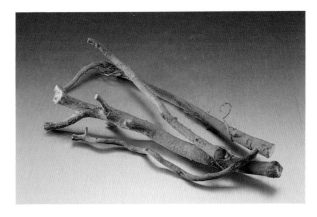

Clerodendrum paniculatum Linn.

VERBENACEAE

Thai name นมสวรรค์, Nom Sawan

Erect shrub, up to 3 m high. Leaves simple, opposite, ovate, 7-38 cm wide, 4-40 cm long, 3-7-lobed. Inflorescence in terminal panicle; calyx campanulate, orange-red; corolla hypocrateriform, orange-red to scarlet. Fruit drupe, globose, greenish blue to black.

Leaf: relieves chest pain. *Flower*: relieves bloody vaginal discharge, inflammation from animal and insect bites, severe infection. *Root*: antimalarial, carminative; relieves tuberculosis, fever. *Stem*: anti-inflammatory for centipede or scorpion bites or carbuncles.

Clerodendrum serratum (Linn.) Moon var. *serratum* Schau.

VERBENACEAE

Thai name อัคคีทวาร, Akkhee Thawaan

Shrub, 1-4 m high. Leaves simple, opposite, oblong, oblong-lanceolate or lanceolate-obovate, 4-6 cm wide, 15-20 cm long, serrate. Flowers in terminal, racemose panicle; corolla 5-lobed, central one dark purplish blue, lateral four light blue. Fruit drupe, subglobose or broadly obovate, turned dark purple or black when ripe.

Dried leaf: pound and take to relieve hemorroids. *Root or stem*: grind, make paste with lime water, apply directly to hemorroids. *Leaf or stem*: decoction as antiflatulent; relieves chronic headache; topically used for ringworm, tinea versicolor, leprosy, joint pain. *Stem*: decoction as antimalarial; relieves abdominal pain. *Wood*: diuretic. *Root*: mix with ginger and coriander, take as antiemetic. *Fruit*: cough remedy.

Clerodendrum infortunatum Gaertn.

VERBENACEAE

Thai name นางแย้มป่า, Naang Yaem Paa

Erect shrub, 0.5-3 m high. Leaves simple, opposite; both surfaces sparsely villous-pubescent, elliptic, broadly elliptic, ovate or elongate ovate, 3.5-20 cm wide, 6-25 cm long, dentate. Inflorescence in terminal, peduncled, few-flowered cyme; flowers white with purplish pink or dull-purple throat, pubescent. Fruit berry, globose, turned bluish-black or black when ripe, enclosed in the red accrescent fruiting-calyx.

Root: diuretic; treatment of intestinal infections and kidney disfunction; boil or grind with water, take to increase milk secretion for post-labor.

Clinacanthus siamensis Brem.

ACANTHACEAE

Thai name ลิ้นงูเห่า, **Lin Nguu Hao**

Scandent shrub, 1.5-4 m high, branchlets erect-drooping clambering over other plants. Leaves simple, opposite, lanceolate or lanceolate-oblong, 2.5-4 cm wide, 7-12 cm long. Inflorescence in terminal dense cyme; flowers dull red with green base, tubular, bilabiate. Fruit capsule.

Leaf or Root: crush with water and topically apply to relieve painful swelling or abscesses, inflammation from centipede or scorpion bites.

Coffea canephora Pierre ex Froehner

RUBIACEAE

Robusta Coffee

Thai name กาแฟ, Kaafae

Shrub, 2-4 m high. Leaves simple, opposite, broadly oblong, 8-12 cm wide, 15-20 cm long; stipules interpetiolar. Inflorescence in short peduncled cyme, each 3-5-flowered, several together in both leaf-axils and forming a many-flowered pseudowhorl; corolla white, tubular, fragrant. Fruit drupe, ovoid-globose, turned red when ripe.

Seed: contains many alkaloids including caffeine, used as central nervous system stimulant, cardiac stimulant and diuretic. Clinical trials prove that drinking coffee increases heart rate and blood pressure, may cause heart burn.

75

Colona auriculata (Desv.) Craib

TILIACEAE

Thai name ปอพราน, Po Phraan

Shrub, up to 2 m high. Leaf simple, alternate, oblong or obovate-oblong, 2.5-5 cm wide, 7-20 cm long; softly hairy especially on lower surface. Inflorescence in axillary, 1-3-flowered cyme; flowers yellow with brownish orange spots. Fruit capsule, indehiscent, ovoid, with 5 longitudinal ridges.

Fruit: mix with *Strychnos nux-blanda* stem bark, *Gloriosa superba* rhizome and cooked rice, used as dog poison.

Connarus semidecandrus Jack

CONNARACEAE

Thai name ถอบแถบเครือ, **Thopthaep Khruea**

Scandent shrub, 2-6 m high. Leaves imparipinnate, alternate; leaflets 3-7, elliptic or lanceolate, 2-7 cm wide, 4-20 cm long. Inflorescence in terminal, long-peduncled, many-flowered panicle; flowers white at first then turn brownish white. Fruit follicle, turned orange when ripe; seeds black with orange-yellow aril.

Leaf: decoction; relieves chest pain. *Root*: antipyretic. *Leaf and stem*: laxative, anthelmintic, antipyretic; treatment of parasitic diseases in children.

Cordyline fruticosa Goppert

AGAVACEAE

Thai name หมากผู้หมากเมีย, **Maak Phuu Maak Mia**

Shrub, 2-4 m high, sparingly branched. Leaf simple, alternate, crowded on tops of branches, lanceolate or linear, entirely green, green with yellow or red streaks or blotches or entirely red, 2-10 cm wide, 20-50 cm long. Inflorescence in upper axillary, large, branched panicle; flowers yellowish white tinged with violet or entirely violet. Fruit berry, globose, turned shining red when ripe.

Leaf: treatment of severe infection; boil or soak in water, bathe to relieve fever with skin manifestations such as measles; topically applied for antipruritic.

Costus speciosus （Koen.） J. E. Smith

COSTACEAE

Crape Ginger, Malay Ginger, Spiral Flag
Thai name เอื้องหมายนา, **Ueang Maai Naa**

Rhizomatous herb, found in swamp area, 1.5-3 m high. Leaf simple, alternate, elliptic-oblong or lanceolate, 5-8 cm wide, 15-40 cm long. Inflorescence in terminal, cylindrical, dense raceme; flowers white with yellow or rose hairy center. Fruit capsule, oblong-triangular, turned red when ripe.

Rhizome: crush and apply around the navel to relieve enlargement of abdomen due to water retention; contains diosgenin, precursor for steroid synthesis.

Crateva adansonii DC.
ssp. *trifoliata* (Roxb.) Jacobs

CAPPARACEAE

Thai name กุ่มบก, **Kum Bok**

Shrub 2 m to tree 25 m high. Leaves palmately compound; leaflets 3, elliptic or ovate, 2-6 cm wide, 2-10 cm long. Flowers in terminal or axillary racemose or corymbose inflorescence; flowers white at first and turned yellow later. Fruit berry on a woody gynophore, globose, pericarp red-violet-brownish tinged when dry.

Leaf: decoction; heart tonic, carminative. *Fresh leaf*: pound and apply to relieve skin diseases, ringworm, tinea versicolor. *Root*: element tonic. *Stem bark*: decoction; carminative, antihiccup. *Wood*: treatment of hemorrhoids; tonic for skinny with yellowish skin; stimulates appetite; tonic for longevity.

Crateva religiosa Ham.

CAPPARACEAE

Thai name กุ่มน้ำ **Kum Nam**

Tree, up to 20 m high. Leaves palmately compound; leaflets 3, elliptic or ovate, 2.5-7 cm wide, 5.5-16 cm long. Flowers in terminal racemose or corymbose inflorescence; flowers white at first and turned yellow later. Fruit berry on a woody gynophore, subglobose to elongate; pericarp greyish when dry.

Leaf: diaphoretic. *Flower*: relieves sore throat. *Fruit*: antipyretic. *Stem bark*: decoction; carminative, antihiccup. *Outer bark*: treatment of hemorroids. *Wood*: relieves dysuria with urinary stones. *Root*: soak in water, take as appetizer. Leaf and stem contain toxic compound, hydrogen cyanide, better taken as pickled vegetable or boiled first.

Cratoxylum formosum (Jack) Dyer ssp. *pruniflorum* Gogel.

CLUSIACEAE

Thai name ติ้วขน Tiu Khon

Tree, 8-15 m high; sap yellowish. Leaves simple, opposite, elliptic-obovate or oblong, 2.5-4.5 cm wide, 3-13 cm long; both surfaces pubescent. Inflorescence in small clusters, above the axils of leaf-scars; flowers light pink. Fruit capsule, ovate-ellipsoid.

Root: decoction; diuretic for dysuria.

Croton crassifolius Giesel

EUPHORBIACEAE

Thai name พังคี, **Phang Khee**

Undershrub, 20-40 cm high. Leaf simple, alternate, elliptic or ovate-lanceolate, 3-5 cm wide, 6-10 cm long, pubescent on both surfaces. Inflorescence in terminal raceme, unisexual, monoecious; flowers yellowish white, Fruit capsule, globose, 3-lobed.

Root: combine with *Clausena harmandiana* root, boil and take as antiflatulent.

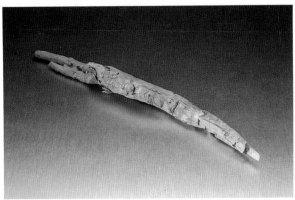

Croton oblongifolius Roxb.

EUPHORBIACEAE

Thai name เปล้าใหญ่, Plao Yai

Small tree, about 8 m high; young leaves brownish. Leaf simple, alternate, oblong, elliptic-oblong, ovate or lanceolate, 5-10 cm wide, 9-30 cm long. Inflorescence in terminal raceme or panicle, unisexual, monoecious or dioecious; flowers greenish yellow. Fruit 3-valved capsule.

Always used together with *C. sublyratus*. *Leaf*: element tonic. *Flower*: anthelmintic. *Fruit*: macerate with 28-40% alcohol, as oxytocic for post-labor. *Stem bark*: promotes digestion. *Stem bark or leaf*: antidiarrheal, blood tonic. *Wood*: anthelmintic. *Root*: carminative; treatment of allergic dermatitis; decoction, used to relieve abdominal pains, bloody and mucous stool. *Stem*: an ingredient in formula for muscular pain. *Leaf*: boil and bathe for pruritic rash.

Cryptolepis buchanani Roem. et Schult.

PERIPLOCACEAE

Thai name เถาเอ็นอ่อน, Thao En On

Woody twiner, lacticiferous. Leaves simple, opposite, ovate or obovate, 3-8 cm wide, 5-18 cm long. Inflorescence in peduncled, axillary cyme; corolla yellowish-white. Fruit a pair of follicles, ovate-lanceolate; seeds with silky comas.

Leaf or stem: strengthen tendons; relieves muscular pain.

Cucurbita moschata Decne.

CUCURBITACEAE

Pumpkin, Cushaw, Winter Squash
Thai name ฟักทอง, **Fak Thong**

Annual creeping herb, 3-10 m long. Leaf simple, alternate, broadly ovate or reniformorbicular, 10-35 cm wide, 7-35 cm long, 5-7-lobed, whitish pilose. Flower solitary, axillary, unisexual, monoecious; corolla campanulate, bright yellow or orange-yellow. Fruit pepo, compressed-globose; pulp yellow or orange; seeds ovate, sordidly white or pale yellow.

Kernel from mature seed: contains cucurbitin; anthelmintic, crush 60 g with milk or water, divide into 3 doses, take every 2 hours followed by castor oil 30 minutes after the last dose; crush with small amount of water, apply to treat mild skin infections and inflammations. *Root*: tonic, aphrodisiac. Clinical trials confirmed that seed extract can kill tapeworm and blood flukes.

Cyathostemma micranthum (A. DC.) J. Sincl.

ANNONACEAE

Thai name นมแมว, **Nom Maeo**

Climber; young branches rusty-brown-tomentose. Leaf simple, alternate, oblong-lanceolate, 2-3.5 cm wide, 6-14 cm long. Inflorescence in terminal or axillary, 2-5-flowered cyme; corolla greenish-yellow, tomentose. Fruit aggregate, globose, turned yellow when ripe.

Root: grind with small amount of water and topically apply for insect bites.

Cyathula prostrata (Linn.) Bl.

AMARANTHACEAE

Thai name พันธุ์งูแดง, **Phan Nguu Daeng**

Perennial herb, 30-50 cm high; stem obtusely quadrangular, densely clothed with fine hairs. Leaves simple, opposite, rhomboid-obovate or rhomboid-oblong, 1-2 cm wide, 3-6 cm long. Inflorescence in terminal or axillary raceme; flower dull pale green. Fruit glabrous utricle.

Stem: diuretic; increase menstrual discharge. *Leaf*: treatment of throat irritations. *Flower*: expectorant. *Root*: relieves abnormal and frequent urination.

89

Cycas pectinata Griff.

CYCADACEAE

Thai name ปรงเขา, **Prong Khao**

Tree, 2-6 m high; trunk often branched. Leaves simple pinnate, crowded at the top, 1.5-2 m long; leaflets narrow linear, 0.5-1 cm wide, 17-25 cm long; petiole with a few small distant spines. Male cone cylindric ovoid to spindle-shaped, about 15 cm wide, 45 cm long. Microsporophylls deltoid-clavate, about 2.5 cm wide, 4 cm long. Macrosporophylls densely tawny silky throughout, about 16 cm long; sterile part triangular-cordate, margin deeply pectinate with subulate teeth about 2 cm long. Seed ovoid, about 4 cm long, orange or yellow-orange.

Sporophyll: tonic, carminative; treatment of biliary diseases, jaundice, bloody sputum or stool. *Root*: grind with rice whiskey or water, topically apply to heal wounds, bruises, swelling and chronic or infected wounds.

90

Cymbidium aloifolium (Linn.) Swartz

(*C. simulans* Rolfe)

ORCHIDACEAE

Thai name กะเรกะร่อน, Kare Karon

Epiphytic orchid with short elongated pseudobulbs. Leaf simple, alternate, rather long and narrow, succulent, about 1.5 cm wide, 45 cm long. Inflorescence in pendulous raceme from near base of pseudobulb; lobes of lip streaked with purple. Fruit capsule with very small seeds.

Fresh leaf: heat and squeeze juice into the ear to treat otitis media.

Decaschistia parviflora **Kurz**

MALVACEAE

Thai name ทองพันดุล, **Thong Phan Dun**

Undershrub, 0.5-1 m high. Leaf simple, alternate, oblong-linear-lanceolate, 2-3 cm wide, 7-10 cm long. Flower solitary, axillary; corolla pale pink, pink, orange or red with whitish center. Fruit capsule, subglobose, brownish hairy.

Fresh root: crush and apply to relieve sprains in pig's legs.

Dendrobium draconis Reichb. f.

ORCHIDACEAE

Thai name เอื้องเงิน, Ueang Ngoen

Epiphytic orchid; stems fairly thick, often long, erect, leafy throughout, 30-50 cm long. Leaf simple, alternate, oblong or elliptic-oblong, about 1 cm wide, 3-4 cm long. Inflorescence short-stalked, axillary, few-flowered; flowers white with orange-red throat, fragrant. Fruit obovate capsule.

Stem: an ingredient in antipyretic formula.

93

Dendrobium trigonopus Reichb. f.

ORCHIDACEAE

Thai name เอื้องคำเหลี่ยม, Ueang Kham Liam

Epiphytic orchid; stems thick, erect, 15-25 cm long. Leaf simple, alternate, oblong, 1.5-2 cm wide, 4-6 cm long. Inflorescence axillary; flowers yellow, fragrant. Fruit capsule.

Stem: an ingredient in antipyretic formula.

Dendrolobium lanceolatum (Dunn.) Schindl.

FABACEAE

Thai name แกลบหนู, **Klaep Nuu**

Erect shrub, 1-3 m high. Leaves imparipinnate, alternate; leaflets 3, ovate or elliptic, 1-2 cm wide, 2.5-5 cm long. Inflorescence in axillary raceme; flowers pea-shaped, pale yellow. Pod flat, 1-seeded.

Root: decoction; treatment of kidney disorders with mucous, yellow or red urine.

Dendrolobium thorelii (Gagnep.) Schindl.

FABACEAE

Thai name ข้าวไหม้, Khaao Mai

Erect shrub, 1-3 m high, all parts long silky pubescent. Leaves imparipinnate, alternate; leaflets 3, ovate, 5-8 cm wide, 10-15 cm long, whitish pubescent on both surfaces. Inflorescence in terminal, dense panicle; flowers pea-shaped, white. Pod flat, curved, whitish pubescent.

Root: decoction; treatment of leukorrhea.

Dendrophthoe pentandra Miq.

LORANTHACEAE

Thai name กาฝากมะม่วง, **Kaafaak Ma Muang**

Epiphytic parasite. Leaf simple, alternate, elliptic-oblong or lanceolate, 2-4 cm wide, 5-8 cm long. Inflorescence in axillary short raceme or fascicle; flowers yellowish green. Fruit berry, oblong-ovoid, turn pinkish red when ripe.

Dried whole plant: infusion; relieves hypertension. Water extract of leaf at high, nearly toxic dose can reduce blood pressure in rats but less activity obtained from methanol extract. More research are needed.

Desmodium gangeticum DC.

FABACEAE

Thai name อีเหนียว, Ee Nieo

Erect shrub, 30-100 cm high; branches puberulous. Leaf unifoliolate, alternate, ovate-oval-oblong, 2-5 cm wide, 3-10 cm long, greyish green pubescent beneath. Inflorescence in axillary raceme; flowers pea-shaped, violet-red. Pod flat, curved, 5-8-jointed.

Root: diuretic; treatment of roundworm in children. *Fresh leaf*: crush and topically apply to treat inflammation from dog bites.

Dianella ensifolia (Linn.) DC.

LILIACEAE

Thai name หญ้าหนูต้น, **Yaa Nuu Ton**

Herb, with subterranean rhizome, 30-60 cm high. Leaf simple, equitant, linear, 1-4 cm wide, 20-60 cm long, sheathed, Inflorescence in terminal panicle; perianth yellowish white. Fruit berry, globose, turned violet-blue when ripe.

Whole plant: an ingredient in the formula for treatment of chronic infected skin diseases.

99

Dioscorea hispida Dennst.

DIOSCOREACEAE

Thai name กลอย, Kloi

Climber; stem thorny; tuber large. Leaves compound, alternate; leaflets 3; middle leaflet elliptic or elliptic-oblong, 6-15 cm wide, 8-25 cm long; lateral leaflets ovate or obovate. Inflorescence unisexual, dioecious, axillary, pendulous; male inflorescence up to 40 cm long, twice or trice compounded; female inflorescence solitary; perianth yellow. Fruit 3-winged capsule.

Tuber: relieves abdominal spasms and colic; fry in vegetable oil, topically apply to remove pus from wounds, clears melasma. Toxic substances such as dioscorine were found in tubers which cause palpitations, nausea, vomiting, throat irritation, sweating, blurred vision and unconsciousness.

Diospyros decandra Lour.

EBENACEAE

Thai name จัน, Chan

Tree, up to 20 m high; young twig villous. Leaf simple, alternate, oblong or elliptic, 2.5-3 cm wide, 7-10 cm long. Flowers unisexual, monoecious. Male flowers cymose; corolla urceolate, whitish. Female flower solitary; corolla as in male flowers but larger. Fruit berry, globose or strongly compressed at both ends, turned yellow and fragrant when ripe; calyx persistent.

Wood: an ingredient in the antipyretic formula.

Diospyros rhodocalyx Kurz

EBENACEAE

Ebony

Thai name ตะโกนา, **Tako Naa**

Tree, up to 15 m high. Leaf simple, alternate, obovate or elliptic, 2.5-7 cm wide, 3-12 cm long. Flower unisexual, monoecious. Male flowers cymose; corolla urceolate, whitish. Female flower solitary; corolla as in male flowers but larger. Fruit globose berry with persistent calyx.

Stem bark and wood: tonic, element tonic; promotes longevity; treatment of impotence, leukorrhea; boil, add salt, used as mouth wash to relieve toothache or gingivitis. *Fruit*: anti-nauseant, antidiarrheal, anthelmintic, anti-inflammatory; treatment of abscesses or chronic infected wounds. *Fruit rind charcoal*: diuretic; soak in water and drink for treatment of leukorrhea. *Bark*: an ingredient in the formula used to promote longevity.

Dipterocarpus obtusifolius Teijsm. ex Miq.

DIPTEROCARPACEAE

Thai name ยางเหียง, **Yaang Hiang**

Deciduous tree, 8-20 m high. Leaf simple, alternate, ovate, 10-20 cm wide, 13-25 cm long. Flowers in axillary, small culster; flowers pinkish. Fruit samaroid, with 2 accrescent sepals.

Leaf: boil, add salt, used as mouth wash to relieve toothache and some dental diseases. *Leaf and latex*: take for infertility. *Tung oil*: expectorant, diuretic; treatment of leukorrhea, uretritis; externally used for wound healing; remove pus from wounds. *Bark*: boil and take as antidiarrheal.

Dischidia major (Vahl) Merr.

(*D. rafflesiana* Wall.)

ASCLEPIADACEAE

Thai name จุกโรหินี, **Chuk Rohinee**

Lacticiferous climber. Leaves simple, opposite, broadly ovate or orbicular, about 2-3 cm wide and long. Flowers in short cymose cluster, 6-8-flowered; corolla urceolate, greenish yellow. Fruit follicle, narrow, curved.

Whole plant: decoction; relieves abdominal pain for patients suffering from peptic ulcer.

Donax grandis Ridl.

MARANTACEAE

Thai name คลุ้ม, Khlum

Erect herb, with slender stem and subterranean rhizome, 0.5-3 m high. Leaf simple, alternate, ovate or oblong-lanceolate, 5-12 cm wide, 10-23 cm long; petiole sheathed. Inflorescence in axillary, pendulous, spicate panicle; corolla yellowish white. Fruit fleshy, globose, turned dirty white when ripe.

Rhizome: antipyretic; treatment of fever with skin manifestations, such as measles, roseolar infantum, rubella, etc.

106

Eclipta prostrata Linn.

ASTERACEAE

Thai name กะเม็ง, Kameng

Annual herb, up to 50 cm high, dull green or red-brown, hispid. Leaves simple, opposite decussate, lanceolate, both surfaces hispid, 1-2.5 cm wide, 3-7 cm long. Inflorescence in axillary head; flowers white. Fruit achene, laterally compressed, black.

Leaf and root: laxative, emetic. *Root*: antiflatulent, liver tonic, spleen tonic, blood tonic; relieves fainting after labor. *Whole plant*: antiasthmatic, antiflatulent, astringent; treatment of bronchitis; treatment of chronic infected skin diseases, ringworm, tinea versicolor. *Juice from stem*: treatment of jaundice.

107

Ellipanthus tomentosus Kurz

ssp. *tomentosus* var. *tomentosus*

CONNARACEAE

Thai name คำรอก, Kham Rok

108

Tree, up to 30 m high, branchlets fulvous-tomentose. Leaf simple, alternate, elliptic, lanceolate or obovate, 3-9 cm wide, 7-22 cm long. Inflorescence in axillary few-flowered glomerulous cluster; flowers white. Fruit dry, dehiscent, ovoid, densely tomentose, brownish when ripe; seeds black with red arilloid.

Branch and stem: antiflatulent, appetizer; relief of abdominal spasm; an ingredient in the antiasthmatic formula. *Stem bark and wood*: decoction; treatment of kidney disfunction with mucous, yellow or red urine.

Enkleia siamensis Nervling

THYMELAEACEAE

Thai name ปอเต่าไห้, Po Tao Hai

Scandent shrub, 2-5 m high; branches brownish puberulous, sometimes transformed into hook-like organs. Leaves simple, opposite, oblong-elliptic or oblong-lanceolate, 1.5-2.5 cm wide, 6-10 cm long. Inflorescence in terminal panicle; flowers bright yellow. Fruit fleshy, subglobose or ovoid.

Wood: expectorant, antipruritic, antiflatulent; treatment of chronic vesicular skin diseases with itching and fever, yaws, leprosy; relieves cough for asthmatic patients with bronchitis. *Root*: decoction; laxative.

Equisetum debile Roxb.

EQUISETACEAE

Horsetail

Thai name หญ้าถอดบ้อง, **Yaa Thot Bong**

Monomorphic pteridophyte, usually growing in marsh; stem with nodes and internodes; surface with 8-25 grooves and ridges, up to more than 1 m high. Leaves sphenophyllous, whorled, green or brown. Cone solitary, terminal on the stems or their branches, oblong.

Stem: decoction; diuretic; combine with *Orthosiphon aristatus* stem and leaf, boil and drink as kidney tonic; treatment of dysuria with urinary stone, leukorrhea.

Erythroxylum cuneatum (Miq.) Kurz

ERYTHROXYLACEAE

Thai name ไกรทอง, Krai Thong

Shrub or small to large tree, 8-40 m high. Leaf simple, alternate, obovate, elliptic or oblong, 2-3 cm wide, 5-11 cm long, stipulate. Inflorescence in axillary cluster; flowers white, light green to yellowish green. Fruit oblong-ovoid drupe, bright red when ripe.

Stem bark: an ingredient in tonic formula; treatment of tendon disorders, muscular pain, beriberi, neuropathy. *Root*: decoction; treatment of kidney disfunction.

Euphorbia hirta Linn.

EUPHORBIACEAE

Thai name น้ำนมราชสีห์, **Nam Nom Raatchasee**

Lacticiferous herb, 15-25 cm high; stems moderately covered with tan hairs. Leaves simple, opposite, upper surface very sparsely puberulous, lower surface sparsely puberulous, 1-1.5 cm wide, 2-4 cm long. Inflorescence axillary, solitary, unisexual, monoecious; flowers aggregate into a cyathium each with one female and several male flowers; cyathial tube green. Fruit 3-lobed capsule, dull light yellow when ripe.

Fresh whole plant: decoction; antitussive, antiasthmatic, antiepileptic; promote lactation; boil with cane sugar and drink as antidysenteric. *Dried whole plant*: infusion; diuretic; for patient with red or unclear urine. Stem extract relieves pain and fever *in vitro*. Clinical trial showed anti-amebic activity.

113

Feronia limonia Swing.

RUTACEAE

Elephant's Apple, Wood Apple, Kavath, Gelingga

***Thai name* มะขวิด, Ma Khwit**

Tree, 6-10 m high. Leaves imparipinnate, alternate; leaflets 5 or 7, less often 3, 6 or 9, oblong-obovate, 0.5-1 cm wide, 1.5-4.5 cm long, pellucid-dotted only along the margin. Inflorescence in terminal or axillary panicle, polygamo-monoecious (unisexual flowers and bisexual flowers); flowers greenish yellow, tinged with red. Fruit berry, globose, turn brownish grey when ripe.

Leaf: antidiarrheal; treatment of bloody vaginal discharge; timing of menstruation; crush and topically apply to treat trauma, abscesses and some skin diseases. Leaf extract can inhibit growth of *Vibrio cholerae in vitro*.

Ficus benjamina Linn.

MORACEAE

Golden Fig

Thai name ไทรย้อยใบแหลม, Sai Yoi Bai Laem

Lacticiferous tree, up to 10 m high, with aerial roots. Leaf simple, alternate, elliptic, lanceolate or ovate-elliptic, 1.5-6 cm wide, 3-12 cm long. Flowers borne in the fig body, axillary, unisexual, monoecious. Fruit fleshy, ellipsoid, ovoid, obovoid or subglobose, turned yellow via orange and dark red.

Aerial root: diuretic; treatment of kidney disfunction with mucous, yellow or red urine; treatment of abnormal and frequent urination.

115

Ficus foveolata Wall.

MORACEAE

Thai name ม้ากระทืบโรง, **Maa Krathuep Rong**

Scandent shrub, lacticiferous, up to 25 m high; young branches, petioles, lower surfaces of the leaves and young receptacles pubescent. Leaf simple, alternate, lanceolate, oblong-lanceolate, ovate or oblong-elliptic, 7-9 cm wide, 12-18 cm long. Inflorescence in axillary, single syconium; flowers unisexual, monoecious, in the same syconium; receptacle globular. Fruit fleshy, globose, reddish inside.

Stem: macerate with rice whisky; drink as tonic, blood tonic.

Ficus racemosa Linn.

MORACEAE

Thai name มะเดื่ออุทุมพร, **Maduea Uthumphon**

Large lacticiferous tree, up to 30 m high; young twigs and figs finely white hairy. Leaf simple, alternate, elliptic, ovate, oblong or lanceolate, 3.5-8.5 cm wide, 6-19 cm long. Flowers borne in fig body, axillary, unisexual, monoecious. Fruit fleshy, ovate-triangular, turned rose-red when ripe.

Root: treatment of fever with skin manifestation, internal fever. *Stem bark*: antidiarrheal (except dysentery and cholera), antiemetic; externally used to stop bleeding; wet dressing for wounds. *Stem*: decoction; relieves muscular pain.

Garcinia schomburgkiana Pierre

CLUSIACEAE

Thai name มะดัน, Ma Dan

Tree, 3-7 m high. Leaves simple, opposite, oblong, lanceolate or oblong-ovate, 2-3 cm wide, 5-8 cm long. Inflorescence in axillary, short cluster, unisexual, monoecious; flowers orange-yellow. Fruit drupaceous, ellipsoid.

Leaf and fruit: macerate with saline water, drink as expectorant, mucolytic; relieves cough; treatment of abnormal menstruation.

Dioecercis erythroclada (Kurz) Tirveng.

(*Gardenia erythroclada* Kurz)

RUBIACEAE

Thai name มะคังแดง, **Ma Khang Daeng**

Tree, 6-12 m high; stems and branches reddish brown; stems aculeate. Leaves simple, opposite, elliptic or obovate, 8-15 cm wide, 15-22 cm long; stipules interpetiolar. Inflorescence in upper axillary cyme; flowers greenish yellow. Fruit fleshy, elongate ellipsoid, with persistent calyx.

Stem: decoction; relieves abdominal pain, improve blood circulation; combine with *Smilax* spp., boil and drink to treat kidney disfunction with mucous, yellow or red urine. *Stem bark*: crush with small amount of water and topically apply to stop bleeding.

Gardenia sootepensis Hutch.

RUBIACEAE

Thai name คำมอกหลวง, **Kham Mok Luang**

Tree, 7-15 m high, with yellowish sap. Leaves simple, opposite, elliptic, obovate-oblong or obovate, 12-18 cm wide, 22-30 cm long; stipules interpetiolar. Flower solitary, axillary; corolla yellowish white at first and turned bright yellow later, fragrant. Fruit fleshy, obovate-ellipsoid.

Seed: boil with water, use as shampoo to kill head lice.

Geophila herbacea (Linn.) O. Ktze.

RUBIACEAE

Thai name มะลิดิน, **Mali Din**

Creeping herb, 10-20 cm high. Leaves simple, opposite, cordate-orbicular or cordate-reniform, 1.5-4.5 cm wide and long; stipules reniform, interpetiolar. Inflorescence in terminal, peduncled, capituliform, few-flowered umbel; corolla-tube tubular-infundibuliform, white. Fruit drupe, globose or broadly ellipsoid, bright red.

Whole plant: decoction; diuretic; decreases internal heat.

121

Glochidion lanceolarium (Roxb.) Voigt

EUPHORBIACEAE

Thai name แดงน้ำ, Daeng Nam

Tree, 8-15 m high. Leaf simple, alternate, oblong or oblong-lanceolate, 3-5 cm wide, 8-12 cm long. Inflorescence in axillary cluster; flowers apetalous; calyx pale green. Fruit schizocarp, longitudinally grooved.

Stem bark: grind with small amount of water and topically apply as antipruritic.

Glycosmis pentaphylla Corr.

RUTACEAE

Thai name เขยตาย, **Khoei Taai**

Shrub, 2-4 m high. Leaves imparipinnate, alternate; leaflets 1-5-foliolate, oblong-elliptic, obovate or lanceolate, 3-5 cm wide, 8-14 cm long. Inflorescence in axillary panicle; flowers white. Fruit berry, globose, white or pink.

Root: grind with water, drink and locally apply to affected area to treat snake bite. Caution: no scientific prove to support this claim. It should be considered first aid before taking the patient to hospital.

123

Gomphia serrata (Gaertn.) Kanis

OCHNACEAE

Thai name ช้างน้าว, Chaang Naao

Tree, up to 25 m high. Leaf simple, alternate, ovate or obovate-lanceolate, 2-6 cm wide, 6-20 cm long. Flowers in terminal or axillary, few-flowered cluster; corolla yellow. Fruit obovoid drupe, turned dark purple or blue-black when ripe.

Root: decoction; antidiabetic; treatment of food poisoning.

124

Grammatophyllum speciosum Bl.

ORCHIDACEAE

Letter Plant

Thai name ว่านเพชรหึง, **Waan Phetchahueng**

Epiphytic orchid; stems up to 3 m high and turned yellow when old. Leaf simple, alternate, relatively long and narrow, curved downwards, about 3 cm wide, 50-60 cm long. Inflorescence in drooping raceme, up to 2 m long with many flowers; flowers pale greenish yellow with large dull orange-brown spots. Fruit capsule.

Stem: crush with rice whisky, filter, drink the filtrate and topically apply the residue to relieve inflammation from snake, scorpion, or centipede poisoning. Caution: no scientific prove to support this claim. It should be considered first aid before taking the patient to a hospital.

125

Graptophyllum pictum Griff.

ACANTHACEAE

Carricature Plant

Thai name ใบเงิน, **Bai Ngoen**

Erect shrub, 1-2 m high. Leaves simple, opposite, elliptic, undulate, variegated by yellowish, whitish or brown-pinkish patches, 4-6 cm wide, 8-12 cm long. Inflorescence in terminal cluster; corolla-tube bilabiate, purplish red. Fruit capsule, elliptic.

Leaf: antipyretic; detoxifying agents, protect liver from poison.

Hedychium coronarium Roem.

ZINGIBERACEAE

Butterfly Lily, Garland Flower, Ginger Lily, White Ginger

Thai name มหาหงส์, Mahaahong

Rhizomatous herb, 1-1.5 m high. Leaf simple, alternate, oblong or lanceolate, 5-8 cm wide, 16-25 cm long. Inflorescence in terminal spike; flowers white, fragrant. Fruit 3-valved, globose capsule.

Dried rhizome: tonic, kidney tonic; grind to fine powder and make pills with honey, for patients suffering from abnormal deterioration of the body, fatigue, pale and muscular pain. *Oil from fresh rhizome*: insecticide.

127

Hegnera obcordata (Miq.) Schindl.

FABACEAE

Thai name ปีกนกแอ่น, Peek Nok Aen

Undershrub, 1.5-2 m high. Leaves pinnately trifoliolate, alternate; terminal leaflet much wider than long, butterfly-shaped, 3-9 cm wide, 1.5-4.5 cm long, lateral leaflets much smaller, finely long-hairy on both surfaces. Inflorescence in terminal and axillary raceme; flowers pea-shaped, violet pink. Pod flat, articles 1-5, ususlly 2-3, horseshoe-shaped.

Root: decoction; improves blood circulation.

128

Helicteres angustifolia Linn.

STERCULIACEAE
Thai name ขี้ตุ่น, Khee Tun

Erect shrub, 1-2 m high; branches brownish tomentose. Leaf simple, alternate, oblong-lanceolate, 1-3 cm wide, 4-8 cm long. Flowers in axillary raceme; flowers pale purple. Fruit capsule, oblong, covered with brownish hairs.

Whole plant: decoction; relieves abdominal spasm. *Root*: treatment of the lower abdominal pain with cloudy, red or yellow urine; boil with water and bathe to heal diabetic wounds.

Naringi crenulata (Roxb.) Nicolson

(*Hesperethusa crenulata* (Roxb.) Roem.)

RUTACEAE

Thai name กระแจะ, **Kra Chae**

Small tree, 3-8 m high, thorny. Leaves imparipinnate, alternate; leaflets 4-13, oval-obovate, 1.5-3 cm wide, 2-7 cm long; petioles winged. Inflorescence in axillary raceme; flowers white. Fruit berry, globose.

Leaf: antiepileptic. *Root*: laxative. *Fruit*: tonic. *Wood*: macerate with rice whisky; tonic, blood tonic; reduce internal heat. *Stem*: treatment of chronic vesicular skin diseases with itching and fever; reduces internal heat; boil and drink to relieve joint, muscular and tendon pain.

Hibiscus mutabilis Linn.

MALVACEAE

Thai name พุดตาน, Phut Taan

Shrub, 2-4 m high; all parts hairy. Leaf simple, alternate, cordate, palmately 3-5-lobed, about 10 cm wide and long; both surfaces densely grayish-stellate-pubescent. Flower solitary, axillary; corolla white at first and turned red via pink. Fruit capsule, globose.

Root: boil and drink or grind with water and topically apply to treat chronic vesicular skin diseases with itching and fever and other skin allergies.

131

Hiptage benghalensis Kurz

MALPIGHIACEAE

Thai name โนรา, Noraa

Large scandent shrub, up to 30 m high. Leaves simple, opposite, elliptic or elliptic-oblong, 4-6 cm wide, 10-15 cm long; young leaf pinkish. Flowers in terminal or axillary raceme-like inflorescence, corolla pink with yellowish blotches, very fragrant. Fruit samara with 3 lateral wings.

Wood: antiflatulent, carminative, aphrodisiac, appetizer; relieves fatigue; promotes longevity.

Hopea odorata Roxb.

DIPTEROCARPACEAE

Iron Wood

Thai name ตะเคียนทอง, **Ta Khian Thong**

Tree, 20-40 m high; young parts hairy. Leaf simple, alternate, ovate-lanceolate, 3-6 cm wide, 8-15 cm long. Inflorescence in axillary panicle; flowers brownish yellow. Fruit samaroid, globular or ovate, with 2 longer wings and 3 shorter wings.

Wood: expectorant; treatment of yaws, blood disorders, fever. *Dried stem latex*: grind and topically apply to heal wounds.

Hydnophytum formicarum Jack

RUBIACEAE

Thai name หัวร้อยรู **Hua Roi Ruu**

Epiphyte, 25-60 cm high. Leaf simple, alternate, elliptic or broadly lanceolate, 2-7 cm wide, 4-15 cm long; stipules interpetiolar. Flowers a few together in shallow, cup-shaped cavities in strongly thickened nodes of the stem; corolla white. Fruit drupe, obovoid, turned orange when ripe.

Stem: grind to fine powder and taken as anthelmintic, heart tonic; treatment of some bone diseases, chronic vesicular skin diseases with itching and fever, knee pain, sprained ankle; an ingredient in antidiabetic formula; treatment of some lung diseases.

Hyptis suaveolens (Linn.) Poit.

LAMIACEAE

Thai name แมงลักคา, **Maeng Lak Khaa**

Erect, branched, glandular-sticky, pleasantly-smelling, annual herb, up to 1.5 m high; stems 4-angled. Leaves simple, opposite decussate, ovate, upper surface finely pilose and with minute, globose, granular glands, lowerside villous and with similar granular glands, 2-5 cm wide, 2.5-6 cm long. Inflorescence in terminal and axillary spike-like raceme with whorls of mostly 4 flowers at each node; corolla tube light lilac with whitish throat and base. Fruit nut, elliptic, flat, black.

Branch and leaf: pound and put in chicken coop as lice repellent.

Ipomoea quamoclit Linn.

CONVOLVULACEAE

Cypress Vine

Thai name คอนสวรรค์, **Khon Sawan**

Annual twiner. Leaf simple, alternate, ovate or oblong in outline, pinnatipartite to the midrib, 1-6 cm wide, 2-10 cm long. Inflorescence axillary, cymose, 2-6-flowered; flowers red. Fruit ovoid capsule.

Stem and leaf: docoction; treatment of bloody cough.

Jasminum rottlerianum Wall. ex DC.

OLEACEAE

Thai name มะลิซาไก, **Mali Saakai**

Climber or reclining shrub, up to 2 m high. Leaves simple, opposite, elliptic, oblong or ovate, 2.5-3.5 cm wide, 5-7 cm long, acrodomatia in the axils of primary vein below. Flowers in terminal subcapitate cymose cluster, 7-13-flowered; corolla white, hypocrateriform. Fruit not developed in Thailand.

Root: female contraceptive.

137

Lasia spinosa Thw.

ARACEAE

Thai name ผักหนาม, Phak Naam

Perennial herb; stems spiny, creeping and upturning; petiole up to 1 m long, aculeate. Leaf simple, alternate, sagittate, pinnatifid, with aculei along veins on lower surface; anterior lobe 35x45 cm, posterior lobes 10x25 cm. Inflorescence in elongated spadix; peduncle up to 75 cm long, aculeate; spathe greenish brown to purplish, up to 55 cm long, slightly twisted; flowers pinkish and finally greenish tan. Fruit leathery berry.

Stem: antitussive, expectorant; boil and bathe to relieve itching from roseolar infantum, measles, rubella and other skin diseases. *Leaf*: relief of abdominal pain. Hydrogen cyanide, poisonous compound, was found in fresh leaf and petiole, better taken as pickled vegetable or boil first.

Laurentia longiflora Peterm.

CAMPANULACEAE

Thai name ปีบฝรั่ง, **Peep Farang**

Erect, lacticiferous, perennial herb, 20-60 cm high. Leaf simple, alternate, lanceolate-obovate, 1.5-4 cm wide, 5-17 cm long, pubescent. Flower solitary, axillary; corolla-tube long, white, fragrant. Fruit capsule, nodding, campanulate.

Leaf: counterirritant for muscular pain, relief of toothache. Caution: Watery juice from fresh plant can irritate soft tissues, especially mouth, throat and eyes.

139

Leea indica Merr.

LEEACEAE

Thai name กะตังใบ. Katangbai

Shrub, 2-4 m high. Leaves imparipinnate, alternate; leaflets oblong, 4-6 cm wide, 12-18 cm long; stipules forming a sheath. Inflorescence in leaf-opposed, compound corymb; corolla light green. Fruit compressed-globose, turned dark red or black when ripe.

Root: antipyretic, diaphoretic; relief of muscular pain; an ingredient in the preparation to treat leukorrhea, intestinal cancer, uterus cancer.

140

Leea macrophylla Roxb. ex Hornem.

LEEACEAE

Thai name พญารากหล่อ, Phayaa Raak Lo

Shrub, 2-4 m high. Leaves imparipinnate, very large, alternate; leaflets broadly ovate, 25-30 cm wide, 30-50 cm long, pubescent; stipules forming a sheath. Inflorescence in axillary cyme; flowers light green. Fruit berry, subglobose and flat at the top, turned black when ripe.

Root: contains red coloring agent for cloth.

141

Leonurus sibiricus Linn.

LAMIACEAE

Motherworth

Thai name กัญชาเทศ, **Kanchaa Thet**

Annual herb, 60-180 cm high. Leaves simple, opposite, elliptic in outline, pinnatifid, 4-5 cm wide, 5-7 cm long. Inflorescence in terminal verticillaster; flowers pinkish violet, bilabiate. Fruit nutlets.

Aerial part: decoction; antimalarial.

142

Lepionurus sylvestris Bl.

OPILIACEAE

Thai name หมากหมก, **Maak Mok**

Shrub, usually less than 1 m high, rarely up to 4 m. Leaf simple, alternate, ovate, oblong, lanceolate or obovate, 3-7 cm wide, 10-16 cm long. Inflorescence in axillary raceme; bracts broadly ovate; perianth yellowish. Fruit ellipsoid or obovoid drupe, turned bright red when ripe.

Root: decoction; relieves some muscular disorder such as muscular pain.

Lepisanthes rubiginosa (Roxb.) Leenh.

SAPINDACEAE

Thai name มะหวด, **Ma Huat**

Tree, 5-10 m high; branches puberulous. Leaves paripinnate; leaflets oblong or oblong-ovate, 3-6 cm wide, 8-12 cm long, puberulous on both surfaces. Inflorescence in terminal panicle; flowers small, whitish. Fruit berry, ovoid, turn dark purple when ripe.

Root: treatment of internal abscesses; grind to fine powder, externally used as antipruritic; apply locally to forehead for fever, headache; decoction; for children suffering from loss of appetite along with white coated, dry tongue and aphthous ulcers in the mouth and throat. *Seed*: treatment of pertussive, cough with bronchial spasm. *Fruit*: tonic. *Leaf*: food preservative.

144

Lepisanthes senegalensis (Poir.) Leenh.

SAPINDACEAE

Thai name หมาว้อ, Maa Wo

Erect shrub, 30-150 cm high. Leaves bifoliolate, alternate; leaflets oblong, lanceolate or narrowly oblanceolate, 1.5-3 cm wide, 8-12 cm long. Inflorescence in terminal panicle; flowers purplish red. Fruit berry, subglobose, turned dark purple when ripe.

Fruit: tonic. *Root*: decoction; treatment of cerebral malaria, fever with vertigo, chest pain and nosebleed; relieves muscular spasm.

Lespedeza parviflora Kurz

FABACEAE

Thai name ก้นบึ้งเล็ก, Kon Bueng Lek

Annual herb, 30-60 cm high. Leaves pinnately trifoliolate, alternate; leaflets elliptic, elliptic-oblong, 2-3 cm wide 4-6 cm long. Inflorescence in terminal panicle; flowers pea-shaped, pinkish. Pod flat.

Root: decoction; antipyretic; reduces internal heat.

Lindenbergia philippensis Benth.

SCROPHULARIACEAE

Thai name หญ้าน้ำดับไฟ, **Yaa Nam Dap Fai**

Perennial herb, 0.5-1 m high. Leaf simple, alternate, ovate or oblong, 1-3.5 cm wide, 2-8 cm long. Inflorescence in terminal spike-like raceme; flowers yellow, bilabiate. Fruit capsule, ovoid.

Fresh whole plant: crush with small amount of rice whisky and apply to the forehead for headache or common colds in children, topically apply to relieve pain and inflammation from burns or abscesses; pound and squeeze for juice, topically apply for vesicular skin infections.

Linociera macrophylla Wall.

(*L. ramiflora* Wall. ex G. Don)

OLEACEAE

Thai name อวบดำ, Uap Dam

Erect shrub, 2-4 m high. Leaves simple, opposite, oblong or elliptic-oblong, 3-5 cm wide, 10-15 cm long. Inflorescence in axillary or ramiflorous panicle; flowers white. Fruit berry, ovoid.

Root: boil with water, used as mouth wash for tooth protection; chew to stop smoking.

Linostoma decandrum (Roxb.) Wall. ex Meissn.

THYMELAEACEAE

Thai name มหาก่าน, **Mahaakaan**

Liana or climbing shrub, up to 5 m high. Leaves simple, opposite, elliptic or ovate-elliptic, 1.5-2.5 cm wide, 3-4 cm long; petioles reddish. Inflorescence in terminal, umbellate panicle; flowers white; calyx brownish red; bracteoles leafy, yellowish green. Fruit fleshy, ovate, with persistent bracteoles.

Stem: rub the stem with a small amount of water on a rough surface for only three times and taken as strong purgative.

149

Litsea glutinosa (Lour.) C. B. Robinson

LAURACEAE

Thai name หมีเหม็น, Mee Men

Tree, 5-15 m high. Leaf simple, alternate, elliptic-oblong or obovate, 5-9 cm wide, 10-20 cm long, pilose beneath. Inflorescence in axillary, umbellate raceme, unisexual, dioecious; perianth 0-3. Fruit berry, globose, turned dark purple when ripe.

Root: relieves muscular pain. *Stem bark*: antidysenteric, antipruritic; treatment of pain from uterine spasm. *Leaf and seed*: crush and topically apply to abscesses to relieve pain. *Stem bark or root bark*: rub on rough surface with dried-immature fruit of *Annona squamosa* and small amount of water, apply to abscesses to drain pus.

Maerua siamensis Kurz

CAPPARACEAE

Thai name แจง, Chaeng

Tree, 5-10 m high; branches glabrous. Leaves palmately compound, alternate; leaflets 3-5, obovate, oblong or linear, 1-3 cm wide, 5-7 cm long. Inflorescence in terminal or lateral corymb or raceme, short terminal panicle or solitary flower in the axils of the upper leaves; petal 0; sepals 4, greenish. Fruit fleshy, ellipsoid or rounded.

Root: tonic, blood tonic, diuretic; treatment of muscular pain, urine disfunction. *Stem bark, root and leaf*: decoction; antimalarial; treatment of jaundice, blurred vision, faintness. *Leaf and young shoot*: pound, used as toothpaste for tooth protection. *Stem*: decoction; tonic; relief of back pain.

Mallotus philippensis (Lam.) Muell. Arg.

EUPHORBIACEAE

Monkey-faced Tree

Thai name คำแสด, **Kham Saet**

Shrub or tree, 5-15 m high; branches brownish pubescent. Leaf simple, alternate, elliptic or ovate, 3-8 cm wide, 6-10 cm long. Inflorescence in terminal or axillary panicle, unisexual, monoecious, Fruit 3-lobed capsule, subglobose, reddish.

Seed: antileprotic, anthelmintic, antipyretic, treatment of vertigo, loss of appetite. *Wood*: decoction; relieves tendon and muscular inflammation, kidney disease with red or yellowish urine. *In vitro and in vivo* researches proved that fruit extract can kill tapeworm.

152

Maytenus marcanii Ding Hou

CELASTRACEAE

Thai name หนามแดง, Naam Daeng

Erect shrub, 3-4 m high; branches long thorny. Leaf simple, alternate, obovate, 2-4 cm wide, 4-9 cm long. Inflorescence in terminal or axillary cluster; flowers yellowish white. Fruit capsule, 3-lobed; seeds brownish.

Wood: element tonic, fat tonic for slim patients. Root: antipyretic.

153

Melaleuca leucadendra Linn. var. *minor* Duthie

MYRTACEAE

Cajuput Tree, Paper Bark Tree, Swamp Tea, Milk Wood

Thai name เสม็ด, Samet

Tree, up to 15 m high; bark papery, whitish, peeling off in layers; branches pendulous. Leaf simple, alternate, elliptic-oblong or lanceolate, 0.5-1 cm wide, 4-8 cm long. Inflorescence in terminal, pendulous spike; flowers in whorls, whitish, with numerous stamens. Fruit capsule, loculicidally dehiscent by 3 valves.

Volatile oil from the leaf: topically apply as counterirritant for sprains, fatigue, pain, or swelling.

Memecylon edule Roxb.

MEMECYLACEAE

Thai name พลองเหมือด, **Phlong Mueat**

Tree, 5-10 m high. Leaves simple, opposite, elliptic-orbicular-oblong, 2-3 cm wide, 3-5 cm long. Inflorescence in axillary, many-flowered panicle; flowers violet; hypanthium campanulate. Fruit berry, globose.

Root: decoction; treatment of chronic vesicular skin diseases with itching and fever. *Stem*: decoction; antiasthmatic. *Stem and leaf*: decoction; diuretic; used for abnormal frequency of urination.

Michelia alba DC.

MAGNOLIACEAE

White Chempaka

Thai name จำปี, Champee

Small to medium-sized tree, 15-22 m high. Leaf simple, alternate, oblong-lanceolate or ovate, 4-9 cm wide, 15-25 cm long. Flower solitary, axillary; perianth white, fragrant. Fruits aggregate, subglobose, rarely fruiting.

Flower: heart tonic, nerve tonic, blood tonic; contains volatile oil, locally apply to relieve headache, swollen eyelids. *Flower and fruit*: element tonic, antipyretic, diuretic; relief of nausea. *Leaf*: antipsychotic; food poisoning. *Stem bark*: antipyretic. *Wood*: menstrual tonic. *Dried root or root bark*: grind with milk, apply to treat abscesses.

Mimosa invisa Mart. ex Colla

FABACEAE

Giant Sensitive Plant

Thai name ไมยราบเครือ, **Maiyaraap Khruea**

Sprawling, straggling, to somewhat climbing, annual, aculeate herb, about 2 m high; stems quadrangular. Leaves bipinnate, alternate, up to 22 cm long; leaflets 6-9 pairs, aculeate, 1-1.5 mm wide, 4-6 mm long. Inflorescence in axillary head; peduncle aculeate; flowers maroonish. Pod flat, straight to slightly curved; margins aculeate; turned light brown when ripe.

Leaf and stem: decoction; diuretic.

Mimosa pudica Linn. var. *hispida* Bren.

FABACEAE

Sensitive Plant

Thai name ไมยราบ Maiyaraap

Straggling, aculeate herb; stems moderately covered with hispid indumentum and sharply pointed, curved spines. Leaves bipinnate, alternate, rapidly sensitive, 10-13 cm long, with 4 pinnae, crowned and appearing digitate, 4.5-8 cm long; leaflets 17-22 opposite pairs, lanceolate, 2.5-3 mm wide, 10-14 mm long, green above, light green and often with dull violet underneath. Inflorescence in axillary head; flowers pink. Pod flat, crowned in capitate infructescences, jointed; turned blackish when ripe.

Whole plant: decoction; diuretic; treatment of kidney disfunction, for anemia, pale, weight loss and muscular pain; *Root*: antidysenteric. Alcoholic extract of the whole plant decreased blood glucose level.

Mitragyna speciosa Korth.

RUBIACEAE

Thai name กระท่อม, Krathom

Tree, 12-15 m high. Leaves simple, opposite, ovate-oblong, 6-9 cm wide, 10-15 cm long; stipules lanceolate, interpetiolar. Inflorescence in terminal panicled head; flowers yellowish white at first then truned brownish yellow. Fruit fleshy, globose.

Leaf: antidysenteric, antidiarrheal; relieves stomachache. Caution: overdose cause vomiting, diarrhea. Some alkaloids show antiprotozoal activity. Alkaloid mitragynine has analgesic activity and people who uptake it can work longer than normal. Considered as a narcotic drug, hallucination and euphoria occur after uptaking. Extended use leads to anorexia, weight loss, darkened and dry skin, psychosis in some cases.

Momordica cochinchinensis (Lour.) Spreng.

CUCURBITACEAE

Thai name ฟักข้าว, Fak Khaao

Tendrilled climber. Leaf simple, alternate, cordate or ovate, 6-15 cm wide and long, 3-5-lobed. Flower solitary, axillary, unisexual, dioecious; flowers yellowish white or yellow with purplish brown center; bracts pubescent. Fruit pepo, oblong, turned red when ripe.

Root: antipyretic, kills head-lice, prevents hair loss. *Fresh leaf*: grind, apply locally for treatment of abscesses. *Leaf*: antipyretic. *Seed*: lung tonic; relief of cough; treatment of tuberculosis, hemorrhoids.

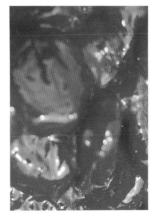

Morinda elliptica Ridl.

RUBIACEAE

Thai name ยอเถื่อน, Yo Thuean

Tree, up to 15 m high. Leaves simple, opposite, elliptic, 5-10 cm wide, 10-12 cm long; stipules interpetiolar. Inflorescence in axillary subglobose head; flowers white. Fruit multiple, subglobose, fleshy, syncarp.

Root: antidiabetic, clothes dyeing for red color. *Wood*: decoction; blood tonic. *Young fruit*: antiemetic. *Ripe fruit*: carminative; stimulate menstruation, *Leaf*: gently heat and apply to chest and abdomen to relieve cough, antiflatulent; apply to hair to kill head-lice.

Morus alba Linn.

MORACEAE

Mulberry Tree, White Mulberry

Thai name หม่อน, Mon

Erect shrub; varieties many. Leaf simple, alternate, ovate, unlobed or lobed, 8-14 cm wide, 12-16 cm long, chartaceous. Flowers in axillary catkin-like inflorescence, unisexual, monoecious; perianth sordidly-white or greenish-white. Fruit multiple, elongate, turned reddish purple when ripe.

Leaf: relief of cough, sedative; decoction, used as an eyebath for treatment of eye infection, blurred vision.

Muehlenbeckia platyclada (F.v. Muell.) Meissn.

POLYGONACEAE

Thai name ตะขาบหิน, Takhaap Hin

Erect shrub, 1-2 m high; branches phyllocladous, strongly branched. Leaf simple, alternate, fugacious, lanceolate-linear, 0.5-1.5 cm wide, 2-5 cm lohg. Inflorescence in axillary cluster, unisexual, dioecious; perianth greenish white. Fruit berry, subglobose, turned red when ripe.

Fresh leaf and stem: grind with a small amount of alcohol or squeeze for the juice, apply to relieve inflammation from centipede or scorpion toxin, treatment of contusion and swelling, painful disorders of muscles.

Mukia maderaspatana Roem.

CUCURBITACEAE

Thai name แตงหนู, Taneng Nuu

Climbing herb; all parts whitish pubescent. Leaf simple, alternate, broadly ovate, 3-5-lobed, 3-7 cm wide, 4-8 cm long, pubescent on both surfaces. Inflorescence in axillary cluster, unisexual. monoecious; flowers yellows. Fruit berry, globose, turned bright red when ripe.

Root: grind and apply to gum to relieve toothache; decoction; antiflatulent. *Stem, young leaf and twig*: taken for treatment of hepatitis, indigestion, nausea, vomiting, bile duct inflammation, bronchitis, asthma. *Seed*: decoction; diaphoretic; seed contains fixed oil, rub to the affected area to relieve muscle pain.

165

Neuropeltis racemosa Wall.

CONVOLVULACEAE

Thai name ม้าทลายโรง, Maa Thalaai Rong

Large woody climber, young branches tomentellous with rusty brown hairs. Leaf simple, alternate, elliptic or elliptic-oblong, 2-7 cm wide, 6-12 cm long. Flowers in axillary racemose cluster, brown-tomentose; corolla broadly campanulate, hairy inside at the base of the filaments; bract ovate or ovate-lanceolate. Fruit capsule, subglobose, with broad-elliptic to orbicular bract.

Wood: macerate in rice whisky, used as tonic, aphrodisiac; increases gastric secretions; decoction; tonic, blood tonic; relieves back and waist pains.

Nicotiana tabacum Linn.

SOLANACEAE

Thai name ยาสูบ, Yaa Suup

Erect, viscid herb, 0.6-2 m high. Leaf simple, alternate, ovate-oblong-lanceolate, 10-20 cm wide, 30-60 cm long, pubescent on both surfaces. Inflorescence in terminal racemose panicle; flowers white or pink, campanulate. Fruit capsule, oblong; seeds very small, dark brown, numerous.

Dried chopped leaf: grind to fine powder and sniff to relieve stuffy nose and common cold. *Root or leaf*: treatment of skin diseases, i.e. scabies, ringworm, tinea versicolor, psoriasis. Leaf contains nicotine, used as insecticide.

167

Olax scandens Roxb.

OLACACEAE

Thai name น้ำใจใคร่, Namchai Khrai

Scandent shrub, up to 5 m high; branches whitish tomentose. Leaf simple, alternate, oblong-lanceolate or oblong-ovate, 2-3 cm wide, 5-7 cm long. Inflorescence in axillary racemose panicle; flowers white. Fruit drupaceous, ovoid.

Stem: decoction; treatment of kidney disfunction.

Oxyceros horridus Lour.
(*Randia siamensis* Craib)
RUBIACEAE
Thai name คัดเค้าเครือ, **Khat Khao Khruea**

Scandent shrub; branches curved aculeate, 3-6 m high. Leaf simple, opposite, elliptic-oblong, 2.5-4 cm wide, 6-10 cm long; stipules interpetiolar. Inflorescence in axillary cymose cluster; flowers white at first and turned yellowish white later, fragrant. Fruit berry, globose.

Fruit: blood tonic, blood purifier; increases menstrual discharge. *Leaf*: soak in water, drink as antipyretic. *Root or Wood*: grind with small amount of water; antipyretic.

Oxystelma secamone (Linn.) Karst.

(*O. esculentum* (Linn.) R. Br.)

ASCLEPIADACEAE

Thai name จมูกปลาหลด, Chamuuk Plaa Lot

Lacticiferous climber. Leaves simple, opposite, linear-lanceolate, 1-2 cm wide, 8-12 cm long. Flowers in axillary cymose cluster or solitary; flowers inside bright pink with purplish lines and brownish blotches and whitish outside. Fruit follicle, lanceolate; seeds brownish with whitish comas.

Root: relief of juandice. *Whole plant*: decoction; used as gargle for sorethroat.

170

Parinari anamensis Hance

ROSACEAE

Thai name มะพอก, **Ma Phok**

Tree, 10-30 m high. Leaf simple, alternate, ovate or elliptic, 4-9 cm wide, 6-15 cm long; white-brownish wooly beneath. Inflorescence in terminal panicle; flowers white. Fruit drupe-like, subglobose or ellipsoid.

Wood: boil with water, drink and bathe to relieve skin diseases; decoction; antiasthmatic. *Stem bark*: wrap with cloth, steam heat and put on the affected area to relieve contusion of internal organs, painful swelling.

Parkia speciosa Hassk.

FABACEAE

Thai name สะตอ Sato

Tree, up to 30 m high; branches puberulous. Leaves bipinnate; pinnae 14-18; leaflets 31-38 pairs per pinnae, linear, 1.8-2.2 mm wide, 6-9 mm long. Inflorescence in terminal, pendulous, pedunculate capitulum, polygamo-monoecious (male flowers and bisexual flowers); flowers yellowish white. Pod flat, 3-5 cm wide, 36-45 cm long, prominently swollen over the seeds.

Seed: carminative; treatment of urination disorders, kidney disfunction.

Peltophorum dasyrachis (Miq.) Kurz ex Baker

FABACEAE

Thai name อะราง, **Araang**

Large tree, up to 30 m high; young shoots brownish-red tomentose. Leaves bipinnate, alternate, 15-40 cm long, stipulate; pinnae 5-9 pairs; leaflets 6-16 pairs, oblong, 5-10 mm wide, 10-15 mm long. Inflorescence in axillary raceme, pendulous, 15-30 cm long; flowers yellow. Pod flat, brownish red; seeds transversely arranged.

Stem bark: treatment of mucous or blood abnormality; decoction; antidysenteric, carminative, antidiarrheal.

173

Phyllanthus acidus (Linn.) Skeels

EUPHORBIACEAE

Star Gooseberry

Thai name มะยม, Ma Yom

Tree, 3-10 m high. Leaf simple, alternate, obliquely ovate or ovate-oblong, 2-4 cm wide, 3.5-8 cm long. Inflorescence in axillary or ramiflorous raceme, unisexual, monoecious; flowers reddish. Fruit drupaceous, compressed-globose, 6-8-lobed, turned pale yellow when ripe, sour.

Root: antipyretic, antipruritic; treatment of skin diseases, allergic dermatitis with watery and pus discharge. *Leaf*: boil and bathe as antipruritic; an ingredient in Ya-Khiew, a formula for antipyretic; remedy for measles, German measles, chicken pox; boil and bathe to relieve fever with skin manifestations, such as measles.

Phyllodium elegans (Lour.) Desv.

FABACEAE

Thai name เกล็ดปลาหมอ, **Klet Plaa Mo**

Erect shrub, 1-2 m high. Leaves imparipinnate, alternate; leaflets 3, terminal leaflet broadly ovate, 3-7 cm wide, 4-9 cm long, lateral leaflets 1.5-4 cm wide, 2-5 cm long. Inflorescence in axillary raceme, covered with two scale-like bracts; flowers white, pea-shaped. Pod flat, linear, articulate.

Root: decoction; relieves liver disfunction; treatment of some psychotic symptoms, including delirium, fibrillation and weight loss, believed to be caused by black magic.

175

Phyllodium longipes (Craib) Schindl.

FABACEAE

Thai name เกล็ดปลา, **Klet Plaa**

Erect shrub, 1.5-2.5 m high. Leaves imparipinnate, alternate; leaflets 3, terminal leaflet ovate-lanceolate, 5-8 cm wide, 8-12 cm long, lateral leaflets 3-4 cm wide, 6-8 cm long. Inflorescence in axillary raceme, covered with two large scale-like bracts; flowers white, pea-shaped. Pod flat, linear, articulate.

Root: decoction; relieves liver disfunction; treatment of some psychotic symptoms, including delirium, fibrillation and weight loss, believed to be caused by black magic.

Phyllodium pulchellum (Linn.) Desv.

FABACEAE

Thai name เกล็ดปลาช่อน, **Klet Plaa Chon**

Erect shrub, 0.6-2 m high. Leaves imparipinnate, alternate; leaflets 3, terminal leaflet oval-oblong or ovate-elliptic, 4-6 cm wide, 6-9 cm long, lateral leaflets 2.5-4 cm wide, 4-6 cm long. Inflorescence in axillary, short raceme, covered with two scale-like bracts; flowers white, pea-shaped. Pod flat, linear, articulate.

Root: decoction; relieves liver disfunction; treatment of some psychotic symptoms, including delirium, fibrillation and weight loss, believed to be caused by black magic.

Physalis minima Linn.

SOLANACEAE

Thai name หญ้าต้อมต้อก, Yaa Tom Tok

Annual herb, up to 50 cm high. Leaf simple, alternate, ovate, 2-3 cm wide, 3-6 cm long. Flower solitary, axillary; corolla greenish yellow. Fruit berry, subglobose, enclosed in the bladdery persistent calyx.

Dried whole plant: decoction; antipyretic, diuretic, antidiarrheal. *Fresh whole plant*: crush with small amount of water, squeeze out the juice and soak with cotton, keep it in mouth, then gently swallow to cure abscesses in throat; grind with water or rice whisky, topically apply to abscesses and swelling.

178

Picrasma javanica Bl.

SIMAROUBACEAE

Thai name กอมขม, **Kom Khom**

Tree, up to 20 m high. Leaves imparipinnate, alternate; leaflets 3-7, elliptic-oblong, 1.5-2.5 cm wide, 4.5-7 cm long. Inflorescence in axillary panicle, unisexual, monoecious; flowers white, yellow or green; calyx persistent. Fruit drupaceous, compressed-globose or ovoid, turned pale yellow when ripe.

Stem bark: antimalarial. Alkaloids from bark show moderate acitvity as central nervous system depressant and antipyretic.

179

Pisonia aculeata Linn.

NYCTAGINACEAE

Thai name คัดเค้าหมู **Khat Khao Muu**

Scandent shrub, up to 8 m high; branches aculeate. Leaf simple, alternate, ovate or elliptic, 3-5 cm wide, 6-9 cm long. Inflorescence in axillary panicle; flowers small, light green. Fruit dry, indehiscent, cylindrical.

Stem: macerate in rice whisky, used as tonic.

Pithecellobium tenue Craib

FABACEAE

Thai name กำลังช้างสาร, **Kam Lang Chaang Saan**

Tree, 5-8 m high; branches thorny. Leaves paripinnate, alternate; leaflets 1-3 pairs, obovate or elliptic, 1.5-2.5 cm wide, 2.5-6 cm long; petiole and rachis winged; stipules thorny. Inflorescence in terminal and axillary head; flowers whitish with numerous whitish stamens. Pod straight to slightly contorted, compressed, indistinctly jointed, slightly constricted between seeds, turned brownish when ripe, dehiscing separately over each joint.

Stem: boil with rice rinse water, take as antipyretic; decoction; treatment of tendon inflammation and waist pain.

181

Polyalthia cerasoides (Roxb.) Benth. ex Bedd.

ANNONACEAE

Thai name กะเจียน, Ka Chian

Medium-sized tree, 12-20 m high. Leaf simple, alternate, lanceolate or elliptic, 2.5-5 cm wide, 6.5-15 cm long. Flowers solitary or 2-3-flowered fascicled; corolla yellowish green, slightly pubescent. Fruit aggregate, ovoid, turned red when ripe.

Root: antipyretic; decoction; for anemia, along with pale, muscular pain and weight loss; as contraceptive for women and tonic for men.

Polyalthia debilis (Pierre) Finet et Gagnep.

ANNONACEAE

Thai name กล้วยเต่า, Kluai Tao

Undershrub, 25-60 cm long; branches brownish tomentose. Leaf simple, alternate, elliptic, elliptic-oblong or oblong-obovate, 2-4 cm wide, 5-8 cm long. Flower solitary, axillary; corolla pale yellow. Fruit aggregate, globose or elongate cylindrical, yellowish, 1-2-seeded, light yellow tomentose.

Root or stem: take fresh or boil and drink for abdominal pain.

184

Polyalthia evecta (Pierre) Finet et Gagnep.

ANNONACEAE

Thai name นมน้อย, Nom Noi

Small shrub, 0.5-1 m high. Leaf simple, alternate, oblong or elliptic, 2-4 cm wide, 6-12 cm long. Flower solitary, axillary; corolla yellow. Fruit aggregate, each spherical, turned reddish brown when ripe.

Root: decoction; relieves abdominal pain caused by muscular cramps; increases milk secretion.

Polyalthia suberosa Thw.

ANNONACEAE

Thai name กลึงกล่อม Klueng Klom

Shrub, 2-4 m high; young twigs rusty-pubescent; stem bark corky ridges, dark red with pink lenticels. Leaf simple, alternate, oblong or oblong-lanceolate, 2-4 cm wide, 5-11 cm long. Flower solitary, extra-axillary; corolla reddish brown outside, yellow inside. Fruit aggregate, each spherical, turned dark purple when ripe.

Leaf and branch contain suberosol which inhibits growth of HIV virus *in vitro*.

Psophocarpus tetragonolobus DC.

FABACEAE

Goa Bean, Manila Pea

Thai name ถั่วพู, Thua Phuu

Twining perennial herb with tuberous main root. Leaves pinnately trifoliolate, alternate; leaflets ovate-rhomboid or ovate-oblong, 4-10 cm wide, 8-18 cm long. Inflorescence in axillary peduncled raceme; flowers pea-shaped, white or blue-violet. Pod with 4 longitudinal wings; seeds white, yellowish brown or black.

Underground stem: tonic; relieves weakness; promotes cheerful sensation.

187

Pterocarpus macrocarpus Kurz

FABACEAE

Thai name ประดู่ป่า, **Pra Duu Paa**

Tree, 15-25 m high. Leaves imparipinnate, alternate; leaflets ovate or ovate-oblong, 3-5 cm wide, 3-9 cm long. Inflorescence in axillary, racemose panicle; flowers pea-shaped, light yellow, fragrant. Fruit samara.

Wood: blood tonic; for patients with anemia, along with pale, muscular pain and weight loss.

Raphistemma hooperianum Decne.

ASCLEPIADACEAE

Thai name ข้าวสารดอกเล็ก, **Khaao Saan Dok Lek**

Lacticiferous, twining glabrous shrub. Leaves simple, opposite, cordate, 7-10 cm wide, 8-12 cm long. Inflorescence in axillary umbelliform cyme; flowers whitish, campanulate. Fruit follicle, fusiform, green; seeds brownish with whitish comas.

Root: emetic; detoxifying agent for food poisioning.

189

Rhodamnia dumetorum (DC.) Merr. et Perry

MYRTACEAE

Thai name พลองแก้มอ้น, Phlong Kaem On

Erect shrub, 1-1.5 m high. Leaves simple, opposite, elliptic or ovate-elliptic, 3-5 cm wide, 6-10 cm long. Flower solitary, axillary; corolla white; stamens numerous. Fruit berry, globose, turned dark purple when ripe.

Root: decoction; antipyretic.

Rothmannia wittii (Craib) Bremek.

(*Randia wittii* Craib)

RUBIACEAE

Thai name หมักม่อ Maak Mo

Tree, 4-10 m high. Leaves simple, opposite, oblong or oblong-elliptic, 3-6 cm wide, 12-16 cm long; stipules interpetiolar. Inflorescence in axillary cymose cluster; corolla campanulate, greenish with whitish margins and purplish blotches. Fruit berry, globose.

Wood or root: decoction; antipyretic. *Stem*: decoction; treatment of venereal diseases.

191

Salacia chinensis Linn.

CELASTRACEAE

Thai name กำแพงเจ็ดชั้น, Kam Phaeng Chetchan

Liana or scandent shrub, up to 5 m high. Leaves simple, opposite, ovate, broadly elliptic, elliptic, elliptic-lanceolate or obovate, 2-4 cm wide, 4-8 cm long. Inflorescence in axillary or ramiflorous panicle; flowers yellowish or yellowish green. Fruit drupaceous, globose or broadly ellipsoid, turned orange-red when ripe.

Stem: decoction; laxative; relieves muscular pain.

Saraca declinata (Jack) Miq.

FABACEAE

Thai name โสกเขา, Sok Khao

Tree, up to 10 m high. Leaves paripinnate, alternate; leaflets 3-7 pairs, ovate or lanceolate, 2-8 cm wide, 8-30 cm long. Inflorescence in terminal or axillary corymb; corolla wanting; calyx orange to red. Pod oblong or lanceolate.

Root: decoction; detoxifying agent for food poisoning.

Sarcostemma brunonianum Wight et Arn.

ASCLEPIADACEAE

Thai name เถาวัลย์ด้วน, Thao Wan Duan

Lacticiferous, trailing, jointed shrub, with pendulous greenish branches. Inflorescence in few-flowered cyme at the nodes; flowers greenish yellow. Fruit follicle, linear; seeds brownish with whitish comas.

Underground stem: tonic; liver, lung and heart tonic.

195

Schima wallichii (DC.) Korth.

THEACEAE

Thai name มังตาน, **Mang Taan**

Tree, up to 25 m high. Leaf simple, alternate, lanceolate, oblong or broadly elliptic, 2-5.5 cm wide, 4.5-13 cm long. Flower solitary, axillary; corolla white; stamens numerous, yellowish. Fruit capsule, subglobose.

Dried flower: macerate in water or infuse, drink to relieve dysuria, convulsion, epilepsy, as soft drink for women after child birth. *Stem and young branch*: relieves nausea; used as eardrop to relieve earache. *Stem bark*: fish poisoning agent; ground into powder, flavoring agent for fragrant incense.

Sindora siamensis Teijsm. ex Miq.
var. *siamensis*

FABACEAE

Thai name มะค่าแต้, Ma Khaa Tae

Tree, above 15 m high; young branches finely pubescent. Leaves paripinnate, alternate; leaflets 3-4-jugate, broadly elliptic or elliptic-oblong, 3-8 cm wide, 6-15 cm long. Inflorescence in terminal or axillary panicle; sepals 4; petal yellowish-red, 1. Pod flat, irregularly ovate, with numerous stout spines.

Stem bark: decoction; treatment of chicken-pox; for weak children who lost of appetite along with white coated dried tongue or aphthous ulcers in the mouth or throat.

Smilax siamensis T. Koyama

SMILACACEAE

Thai name เขืองแดง, **Khueang Daeng**

Large climber, 2-8 m long; stem lightly green. Leaf simple, alternate, elliptic or narrowly elliptic, 3-12 cm wide, 6-20 cm long; wings broadly ovate; tendrils 7-20 cm long. Inflorescence borne on branches, unisexual, dioecious, both sexes with 7-25 umbels; umbels 18-40-flowered; tepals yellow or yellowish green. Fruit baccate, globose.

Rhizome: decoction; liver tonic.

Solanum stramonifolium Jacq.

SOLANACEAE

Thai name มะอึก, **Ma Uek**

Erect shrub, 1-2 m high; all parts pale brownish tomentose. Leaf simple, alternate, broadly ovate or orbicular, lobed, 15-25 cm wide, 20-30 cm long, densely tomentose. Inflorescence in axillary, short cluster; flowers white. Fruit berry, globose, turned brownish yellow when ripe; seeds many.

Fruit or root: mucolytic; cough remedy. *Root*: treatment of fever with skin manifestations, i.e. measles, chickenpox.

Solena heterophylla Lour.

CUCURBITACEAE

Thai name ตำลึงตัวผู้, **Tam Lueng Tua Phuu**

Slender climber with unbranced tendrils. Leaf simple, alternate, triangular-lanceolate to cordate-oblong, 1.5-3 cm wide, 5-10 cm long, margin slightly 3-lobed, chartaceous. Flowers unisexual, monoecious, male inflorescence axillary; female flower solitary; flower white or yellowish white. Fruit berry, oblong.

Root: decoction; laxative.

Sphenodesme involucrata (Presl) Robinson

SYMPHOREMACEAE

Thai name เถาวัลย์ปูน, Thao Wan Puun

Scandent shrub, up to 10 m high. Leaves simple, opposite, elliptic or elliptic-ovate, 5-8 cm wide, 10-14 cm long. Inflorescence in axillary, short cluster; flowers pale yellow; bracts 6-8, papery. Fruit dry, indehiscent.

Stem: grind with small amount of water, take for abdominal pain. Note: not recommended to take more than 4 times.

203

Stachytarpheta indica Vahl

VERBENACEAE

Thai name พันธุ์งูเขียว, **Phan Nguu Khieo**

Spreading herb, 60-120 cm high. Leaves simple, opposite, oblong, elliptic or ovate, 1.5-5 cm wide, 2-8 cm long. Inflorescence in terminal, elongate spike; flowers bluish violet. Fruit schizocarp, linear-oblong.

Whole plant: decoction; diaphoretic, antipyretic, diuretic.

Stemona phyllantha Gagnep.

STEMONACEAE

Thai name สามสิบกีบ, **Sam Sip Keep**

Perennial, climbing herb, up to 10 m long; root fusiform, fascicled; stem greenish, succulent. Leaf simple, alternate, ovate, 2-4 cm wide, 7-10 cm long. Flowers 2-3-fascicled, axillary; corolla light yellow with greenish longitudinal lines and reddish center. Fruit dry, dehiscent.

Root: used as shampoo to kill head-lice.

Stephania venosa (Bl.) Spreng.

MENISPERMACEAE

Thai name กลิ้งกลางดง, **Kling Klaang Dong**

Slender climber, containing reddish sap; leafy stem herbaceous, annual, arising from a large exposed tuber. Leaf simple, alternate, broadly triangular-ovate, margin often slightly lobed, 7-12 cm wide, 6-11 cm long, lower surface minutely papillose. Inflorescence in axillary umbelliform cyme, 4-16 cm long; flowers orange. Female inflorescence much more condensed than the male. Fruit drupe, obovate.

Stem: anthelmintic. *Root*: nerve tonic. *Underground stem*: macerate in rice whisky, drink as tonic, aphrodisiac; dry and grind into powder, make pills with honey for longevity, as appetizer. *Leaf*: treatment of acute or chronic wounds.

Streblus asper Lour.

MORACEAE

Siamese Rough Bush, Tooth Brush Tree

Thai name ข่อย, Khoi

Lacticiferous tree, 5-15 m high. Leaf simple, alternate, oblong-ovate, elliptic or obovate, 2-4 cm wide, 4-8 cm long, very scabrid. Inflorescence unisexual, dioecious; male flowers in axillary head; female flowers in axillary, 2-4-flowered cluster; corolla yellow. Fruit fleshy, ovoid, turned yellow when ripe.

Stem bark: antidiarrheal, antidysenteric, antipyretic; as tooth powder for periodontitis, toothache; treatment of skin diseases, wounds. *Seed*: element tonic, carminative; appetizer; macerate in rice whisky or boil with water and drink for longevity; boil with fixed oil, locally apply to relieve hemorrhoids. *Leaf*: gently heat and infuse as laxative. *Wood*: smoke for treatment of nasal polyposis.

Strophanthus gratus Franch.

APOCYNACEAE

Thai name บานทน, Baan Thon

Erect shrub, 3-5 m high. Leaves simple, opposite, oval-oblong, 4-6 cm wide, 9-13 cm long. Inflorescence in terminal cyme, 5-20-flowered, fragrant; flowers pinkish-violet, campanulate. Fruit follicle, linear; seeds brownish with whitish comas.

Seed: contains cardiac stimulant, ouabain. Note: Considered to be extremely poisonous, not recommended as herbal medicine, must be formulated as injection. Toxic symptoms are nausea, diarrhea, arrythmias. For first aid, allow the patient to vomit immediately and take to hospital.

Strophanthus scandens Roem. et Schult.

APOCYNACEAE

Thai name เครือน่อง, **Khruea Nong**

Lacticiferous liana, up to 12 m high. Leaves simple, opposite, elliptic, obovate or ovate, 3-9 cm wide, 5-15 cm long. Inflorescence in terminal, pedunculate, 2-16-flowered cyme; flowers at first white, turned red via yellow; tails yellow, turned purple via red; corona lobes red, turned purple. Fruit a pair of follicles, divergent at an angle of 180-200, elongate, 2-5 cm wide, 13-30 cm long; seeds brownish with whitish comas.

Latex from stem: mixed with other poisonous latex and used as arrow poison for animal hunting. Toxic symptoms are the same as *Strophanthus gratus*.

Strophioblachia glandulosa Pax
var. *pandurifolia* Airy Shaw

EUPHORBIACEAE

Thai name บ๊าซาด, Baa Saat

Erect undershrub, 25-50 cm high. Leaf simple, alternate, ovate or ovate-oblong, 3-5 cm wide, 6-9 cm long, sparsely pubescent on both surfaces. Inflorescence in terminal and axillary, lax panicle, apetalous; stamens numerous. Fruit capsule, 3-lobed.

Root: decoction; increases milk secretion for post-labor.

Strychnos lucida R. Br.

STRYCHNACEAE

Thai name พญามูลเหล็ก, **Phayaa Muun Lek**

Small tree, 4-10 m high. Leaves simple, opposite, ovate, 2.5-4 cm wide, 4-6 cm long. Inflorescence in axillary cymose panicle; flowers light green. Fruit berry, globose; seeds flat, button-shaped.

Wood: grind with a small amount of water, take to treat all types of fever, uterus or ovarial cancers and lung cancer in men; topically apply as antipruritic, antidandruff; an ingredient in antidiabetic formula. *Seed and stem bark*: tonic; treatment of cholera. *Leaf*: grind to fine powder and apply to relieve bruises and swelling.

Strychnos nux-blanda A. W. Hill

STRYCHNACEAE

Thai name ตูมกาขาว, **Tuumkaa Khaao**

Tree, up to 25 m high. Leaves simple, opposite broadly ovate, 8-10 cm wide, 10-14 cm long. Inflorescence in axillary cymose panicle; flowers greenish yellow. Fruit berry, globose, larger than *S. lucida*, turned orange when ripe; seed flat, button-shaped.

Root: antimalarial; grind with small amount of water, topically apply to snake bite as anti-inflammatory agent; decoction; laxative. *Stem bark*: combine with *Colona auriculata* fruit, *Gloriosa superba* rhizome and food, used as poisoning agent for dogs.

Tacca integrifolia Ker-Gawl.

TACCACEAE

Black Lily

Thai name ว่านพังพอน, **Waan Phangphon**

Herb with cylindrical rhizome, 40-60 cm high. Leaf simple, basal, oblong or lanceolate, 7-17 cm wide, 20-60 cm long. Flowers in umbellate, involucrate, 6-30-flowered cluster; perianth greenish or dark purple; involucral bracts greenish-white or dark purple. Fruit berry-like, 6-angled.

Rhizome: decoction; aphrodisiac; treatment of hypotension.

Tadehagi godefroyanum (O. Ktze.) Ohashi

FABACEAE

Thai name ไชหิน, Chai Hin

Erect shrub, 1.5-3 m high. Leaf unifoliolate, alternate, ovate or ovate-oblong, 5-8 cm wide, 12-15 cm long; blades greyish green; petioles winged. Inflorescence in terminal raceme; flowers pea-shaped, dark reddish purple. Pod flat, articulate.

Root: decoction; antiemetic; treatment of bloody discharge *via* both mouth and anus.

Tectona grandis Linn.

VERBENACEAE

Teak

Thai name สัก, Sak

Large tree, up to 50 m high; young parts tomentose. Leaves simple, opposite, broadly elliptic, 6-50 cm wide, 11-95 cm long. Inflorescence in uppermost leaf-axils and terminal panicle; flowers white, short-hypocrateriform. Fruit drupe, subglobose, densely tomentose, completely enveloped by the enlarged fruiting-calyx.

Wood: diuretic, antipyretic; relieves edema; an ingredient in antidiabetic formula. *Stem bark*: antidiarrheal. *Leaf and wood*: carminative, antidiabetic; treatment of abnormal urination, kidney disfunction. In animal tests, leaf extract decreased glucose blood level with rapid onset and short duration of action.

Terminalia citrina Roxb. ex Flem.

COMBRETACEAE

Thai name สมอดีงู, Samo Dee Nguu

Tree, 20-30 cm high; young branches pubescent. Leaves simple, opposite, subopposite or alternate, elliptic, narrowly elliptic or oblong-elliptic, 2-6 cm wide, 3-14 cm long. Inflorescence in terminal panicle, apetalous; calyx greenish yellow. Fruit drupe-like, ellipsoid, turned greenish purple when ripe.

Young fruit: laxative, antipyretic, carminative; treatment of blood toxemia for post-labor (complications caused by abnormal menstruation and postpartum toxemia).

216

Theobroma cacao Linn.

STERCULIACEAE

Cocoa Tree

Thai name โกโก้, Ko Ko

Tree, 3-8 m high. Leaf simple, alternate, oblong-obovate or oblong, 10-15 cm wide, 15-30 cm long. Flowers solitary or in cluster, cauliflorous or ramiflorous; corolla yellowish white; staminodes dark purple, white-tipped. Fruit fleshy, ovoid-ellipsoid, prominently wrinkled, turned purple or yellow when ripe; seeds ellipsoid, brownish.

Seed: tonic; fixed oil expressed from seed is used as suppository base. Theobromine and caffeine from seeds are central nervous system stimulant and promote urine excretion.

Thottea parviflora Ridl.

ARISTOLOCHIACEAE

Thai name หูหมี, Huu Mee

Shrublet, about 1 m high; branches pubescent. Leaf simple, alternate, ovate, obovate, elliptic or oblong, 7-9.5 cm wide, 16.5-22 cm long. Flowers in axillary racemose cluster; perianth connate, pink or violet. Fruit capsule, elongate, 4-angled, twisted, light brown.

Whole plant: decoction; diuretic; treatment of prostitis.

Thottea tomentosa (Bl.) Ding Hou

ARISTOLOCHIACEAE

Thai name บูดูบูลัง, **Buuduu Buulang**

Undershrub, about 30 cm high, densely tomentose. Leaf simple, alternate, elliptic, oblong, ovate or broadly ovate, 8-11 cm wide, 11-14 cm long. Flowers in racemose cluster near the base of the stem; perianth cup-shaped, purplish-brown outside, yellowish-brown inside. Fruit capsule, elongate, with 4 distinct valves, twisted, brownish.

Whole plant: decoction; diuretic; treatment of prostitis.

Toddalia asiatica (Linn.) Lamk.

RUTACEAE

Thai name เครืองูเห่า, **Khruea Nguu Hao**

Liana, 2-20 m long; branches aculeate. Leaves digitately-trifoliolate, alternate: leaflets oblong-obovate, 1-2.5 cm wide, 3-8 cm long, pellucid-dotted. Inflorescence in terminal and axillary, umbellate panicle; flowers greenish yellow. Fruit hesperidium, globose, turned orange when ripe.

Stem: decoction; diuretic; relieves tendon and waist pain.

Trema orientalis (Linn.) Bl.

ULMACEAE

Peach Cedar

Thai name พังแหรใหญ่, **Phang Rae Yai**

Tree, 5-30 m high. Leaf simple, alternate, oblong or ovate-cordate, 3-5 cm wide, 8-10 cm long; base oblique, inequilateral, scabridulous above, densely canescent beneath. Inflorescence in axillary short cyme, unisexual, monoecious; flowers light green, small. Fruit drupe, ovoid, turned dark purple when ripe.

Stem bark: chew and keep in mouth for 30 minutes to cure chronic ulcer on lips. *Wood or root*: grind with small amount of water, take to relieve internal fever and thirst.

Trevesia palmata Vis.

ARALIACEAE

Thai name ต้างหลวง, Taang Luang

Tree, 3-6 m high. Leaf simple, alternate, crowded at the top, 5-9-lobed, 20-30 cm wide and long, brownish green, brownish puberulous. Inflorescence in ramiflorous umbellate panicle; flowers yellow. Fruit fleshy, obconical.

Young flower: appetizer.

Tribulus terrestris Linn.

ZYGOPHYLLACEAE

Ground Barnut, Small Caltrops

Thai name หนามกระสุน, Naam Krasun

Annual with prostrate, decumbent herb; primary branches up to about 160 cm long, radiating from the crown of the stem, pubescent. Leaves paripinnate, opposite; leaflets oblong, 2-5 mm wide, 5-15 mm long; both surfaces silky hairy; stipules lanceolate or falcate. Flowers solitary or short cymose inflorescence; corolla bright yellow. Fruit capsule, of 5 or fewer tuberculate, woody cocci; each coccus with 2 pairs of sharp stiff spines.

Whole plant: diuretic; treatment of dysuria with urinary stones, leukorrhea, kidney disfunction, abnormal urination; contains high potassium level. Not recommended for patients with heart diseases. Clinical study in healthty volunteers showed that whole plant infusion increases urine excretion.

Trigonostemon reidioides (Kurz) Craib

EUPHORBIACEAE

Thai name โลดทะนง, Lot Thanong

Undershrub, 0.5-1.5 m high; all parts pubescent. Leaf simple, alternate, oblong or oblong-lanceolate, 2-4 cm wide, 7-12 cm long, finely pubescent on both surfaces. Inflorescence in axillary or ramiflorous panicle, unisexual, monoecious; flowers white, pink or purple. Fruit capsule, 3-lobed, subglobose.

Root: grind with water, taken as laxative, antiasthmatic, emetic for food poisoning, especially from mushrooms and shells; treatment of bloody and mucous sputum or stool; topically apply to abscesses, sprains, swelling, bruises, snake bites, especially to treat snake neurotoxin. Needs further investigation.

Typhonium trilobatum Schott

ARACEAE

Thai name อุตพิด, Uttaphit

Perennial herb, 10-45 cm high. Leaf simple, basal, cordate-hastate, unlobed, 3-lobed or 3-partite, 10-20 cm wide and long. Inflorescence in terminal spadix; spathe velvety reddish purple or dark purple, unisexual, monoecious; flowers yellowish. Fruit berry, globose.

Rhizome: treatment of localized abdominal rigidity; boil with fixed oil, locally apply as wound healing or treatment of infected wounds.

Uvaria rufa Bl.

ANNONACEAE

Thai name นมควาย, **Nom Khwaai**

Scandent shrub, up to 5 m high; young branches reddish-brown puberulous. Leaf simple, alternate, elliptic or ovate, reddish-brown puberulous on both surfaces, 2.5-3.5 cm wide, 4.5-10 cm long. Flowers in ramiflorous cluster, 2-3-flowered; corolla dark red, fragrant. Fruit aggregate, ovoid or obovate, turned bright red when ripe.

Wood and root: decoction; treatment of interrupted fever. *Root*: tonic for women to promote recovery of health during postpartum peroid; increases milk secretion. *Fruit*: grind with water, apply to relieve itching

Vanilla planifolia Andr.

ORCHIDACEAE

Vanilla

Thai name วานิลลา, Waaninlaa

Orchidaceous climbing herb; stem succulent, greenish, jointed. Leaf simple, alternate, oblong or lanceolate, 3-6 cm wide, 10-15 cm long, thick, succulent. Inflorescence in terminal or axillary raceme; flowers greenish yellow. Fruit capsule, elongate-linear, after drying strongly scented.

Fruit: contains vanillin, used as flavoring agent for food and drugs.

Vitex glabrata R. Br.

VERBENACEAE

Thai name ไข่เน่า, Khai nao

Medium-sized tree, 10-25 m high. Leaves digitately quinquefoliolate, opposite; leaflets elliptic-obovate or oblanceolate, 4-6 cm wide, 10-13 cm long. Inflorescence in axillary, corymbose cyme; corolla light violet, broadly cylindrical, villous. Fruit drupaceous, ovoid or obovoid, turned dark purple when ripe.

Stem bark: antidiarrheal, antidysenteric, antipruritic; treatment of children disease usually caused by infection of intestinal parasites, common symptoms are anorexia, weight loss, pale and diarrhea. *Root*: antidiarrheal, appetizer.

Volkameria fragrans Vent.

VERBENACEAE

Thai name นางแย้ม, Naang Yaem

Erect shrub, 1.5-3 m high. Leaves simple, opposite, broadly ovate, 8-11 cm wide, 12-16 cm long, moderately clothed with patent hairs. Inflorescence in terminal, dense subcapituliform corymb; flowers white, outermost white with purplish red; calyx purplish red; stamen and ovary absent.

Root: diuretic; treatment of abdominal pain, intestinal diseases, kidney disfunction.

230

Xylia xylocarpa (Roxb.) Taub.
var. *kerrii* (Craib et Hutch.) Nielsen

FABACEAE

Iron Wood

Thai name แดง, Daeng

Tree, 5-30 m high; young branches and shoots yellowish puberulous. Leaves bipinnate, alternate; pinnae 2; leaflets 4-5 pairs per pinna, ovate or ovate-oblong, 3-7 cm wide, 7-20 cm long. Inflorescence in terminal or axillary, single or compound heads; flowers white or yellowish white. Pod flat, kidney-shaped.

Stem bark: antidiarrheal; decoction; used to cure exanthematous fevers. *Wood*: boil with water and drink as laxative; treatment of uterus or ovarial cancer in women and lung cancer in men; *Stem bark and wood*: decoction; take a glass daily for internal injury, bloody vaginal discharge. Not recommended for pregnant women. *Flower*: anti-pyretic, heart tonic.

231

Zingiber ottensii Val.

ZINGIBERACEAE

Thai name ไพลดำ, Phlai Dam

Rhizomatous herb; pseudostem up to 1.5 m high. Leaf simple, alternate, 6-8 cm wide, 26-30 cm long. Inflorescence protuded from underground rhizome near pseudostem; Inflorescence about 9 cm long; bracts greenish red when young and turned pink or red later; corolla light yellow. Fruit capsule, subglobose.

Root and rhizome: grind and squeeze out liquid, apply to sprains, bruises and swelling; treatment of beriberi, muscular pain; increases menstrual discharge. Squeeze juice from rhizome, mix with heated salt, take as mild laxative, antidysenteric, astringent in the intestine.

232

Thai-style Herbal Steam Bath

Thai-style herbal steam bath is a process whereby a mixture of fresh or dried herbs is boiled in water and the steam arising from this process is used first for inhalation and later as a steam bath. The authentic practice derived from the age-old tradition, not only helps to clean the body and the mind but also helps to alleviate certain ailments such as various respiratory complaints, some skin diseases, muscle stress and strains and common colds. Herbal steam bath is specially recommended for woman after childbirth. More recently, it has been employed as part of the rehabilitation programs for drug addicts as well as for weight reduction.

Preparation of the Herbal Ingredients

For the steam bath, it is preferable to use fresh herbs as some of the chemical constituents are usually lost during the drying process, particularly in the case of volatile oil containing plants. These ingredients may be classified into four groups.

Group 1. Aromatic herbs

Group 1. Aromatic herbs

This group consists of herbs containing votatile oils as their principal active ingredients. Apart from their pleasant aroma, these principles may also help to stimulate peripheral circulation, soothe skin conditions and relieve complaints, such as skin rashes, aches and pains and nasal congestion. Most of the drugs in this category are derived from the roots or rhizomes, while others are derived from the leaves, stem and fruit rind. Examples of this group of herbs are shown in the following table.

233

Common / Thai names	Scientific names	Part(s) used	Indications
Turmeric/Khamin Chan	*Curcuma longa*	rhizome	itching and infected wounds
Zedoary/Khamin Oi	*C. zedoaria*	rhizome	wounds
-/Phlai	*Zingiber purpureum*	rhizome	sprains, muscular pain, stuffy
Sweet Flag/Waan Nam	*Acorus calamus*	rhizome	sprains, muscular pain wound healing, colds,
-/Proh Hom	*Kaempferia galanga*	rhizome	relieve common cold, nosebleed
Pomelo/Som O	*Citrus maxima*	leaf	dandruff, dry hair condition
Camphor Tree/Naat	*Blumea balsamifera*	leaf	asthma, rhinitis, diaphoretic
Lemon Grass/Takhrai	*Cymbopogon citratus*	leaf and stem	fungal and bacterial infections
Leech Lime/Ma Kruut	*Citrus hystrix*	fruit rind	dandruff, dry hair condition

Group 2. Herbs possessing a sour taste

This group of herbs is weakly acidic and possesses a mild anti-bacterial activity as well as cleansing property. They are recommended for itching, certain skin diseases and dandruff.

common/Thai names	Scientific names	Part (s) used
Soap Pod/Som Poi	*Acacia concinna*	leaf and pod
Tamarind/Ma Khaam	*Tamarindus indica*	leaf
Leech Lime/Ma Kruut	*Citrus hystrix*	whole fruit

Group 2. Herbs possessing a sour taste

Group 3. Sublimable ingredients

Crystaline compounds, such as camphor and borneo-camphor can sublime and give off fragrance when heated. These ingredients are not boiled with the other ingredients but they are gradually sprinkled onto the boiling concoction so that they can be inhaled with the steam. Both camphor and borneo-camphor have cardiotonic as well as antidermatitis properties.

Group 4. Herbs for specific ailments.

Different herbs may be added to the steam bath for the treatment of certain ailments, especially the herbs that possess the following properties: anti-inflammatory, antihistaminic, antibacterial, antifungal, antipyretic, demulcent, expectorant, decongestant and diaphoretic. Some of these herbs are shown in the table.

Common / Thai names	Scientific names	Part(s) used	Indications
Sea Holly/Ngueak Plaa Mo	*Acanthus ebracteatus*	leaf	abscesses, infected wound
-/Chingchee	*Capparis micracantha*	leaf/stem/root	bronchitis; muscular cramps, swelling vesicular chronic infected skin diseases
Garden Quinine/ Sammangaa	*Clerodendrum inerme*	leaf	treatment of skin diseases, itching
-/Tamlueng	*Coccinia grandis*	leaf/stem	itching, inflammed wounds
-/Phak Bung ruem	*Enydra fluctuans*	leaf/stem	swelling, inflammed wounds
Goat's Foot Creeper/ Phak Bung Thale	*Ipomoea pes-caprae*	leaf	skin inflammation, allergy
-/Phak Naam	*Lasia spinosa*	whole plant	measles and fever accompanied with rashes
-/Phak Chee Lom	*Oenanthe stolonifera*	whole plant	cough remedy, allergic dermatitis with watery or pussy discharge, detoxification by means of sweating
Soapberry/Makham Dee Khwaai	*Sapindus emarginatus*	fruit	seborrhoea, ringworm, fungal infection, dandruff, skin tonic
-/Thong Phan Chang	*Rhinacanthus nasutus*	leaf/root	fungal diseases
Castor Oil Plant/ Lahung	*Ricinus communis*	leaf	itching, and infected wounds

In gereral, there is no restriction as to the number of different herbs being used but all four groups of herbs should be included as a rule. The selection of specific herbs in group 4 will depend on each particular ailment being treated.

Preparation of the Herbal Steam

All the herbal ingredients previously selected are placed into a large covered container and sufficient water is added. The mixture is then boiled for 5-7 minutes. After that the container is removed from the heater and placed in a "kra chome", small tent-like structure, where the patient or customer is waiting. To begin the process, the lid of the container is slightly opened and small amounts of camphor and borneo-camphor are added to the boiling concoction at a time. The patient inhales the herb scented steam and remain seated in "kra chome" which acts like an oven. The patient will

Group 3. Sublimable ingredients

235

be sweaty from the herbal steam. The "kra chome" is then removed when the water in the container becomes lukewarm and the patient uses the herb ladden water to wash his/her hair and body.

The Benefits of Thai-style Herbal Sauna Bath.

In recent years, sauna bath which employs dry heat has become popular in many countries including Japan and Korea. By comparison, Thai-style herbal steam bath which has been practised for hundreds of years seems to offer more benefits which may be summed up as follows:

✴ Clearing the airways throughout the respiratory system: to relieve sinusitis, bronchial asthma.

✴ Circulatory stimulation: it helps to stimulate the blood circulation as well as to ease muscular aches and pains.

✴ Clensing the skin: steam will help to enlarge the pores in the skin, thus assisting the removal of extraneous materials through sweating and the skin washes relieve inflamed skin conditions.

✴ Relaxation: during the 20-30 minutes spent in the steam bath, both the body and mind will become relaxed and will revitalized as the result.

✴ Weight reduction: for those who want to reduce their weights, herbal steam bath offers a safer alternative way to lose the extra pounds without having to resort to taking drugs.

Even though herbal steam bath is considered a safe form of therapy, it is not recommended for people suffering from hypertension or abnormal heart function.

It may be said that herbal steam bath is an integral part of the traditional Thai-style of health care regimen and still plays an important role in postnatal care for women after childbirth, especially in rural areas. With the rising popularity of steam bath in most health clubs around the country, herbal steam bath should be revived and adapted to suit modern style of living, thereby resulting in the preservation of an important part of Thai cultural heritage.

Group 4. Herbs for specific ailments.

236

BIBLIOGRAPHY

กิ่งแก้ว เกษโกวิท และคณะ. หมอพื้นบ้านและการดูแลสุขภาพตนเองของชาวบ้านอีสาน: กรณีศึกษา. มหาวิทยาลัยขอนแก่น, 2536.

โกมาตร จึงเสถียรทรัพย์ และคณะ. โครงการวิจัยเพื่อศึกษาสภาพความนิยมในการรักษาแบบพื้นบ้าน โดยการใช้สมุนไพรของชุมชน ในเขตอำเภอชุมพวง โรงพยาบาลชุมพวง จังหวัดนครราชสีมา, 2529.

โครงการศูนย์ข้อมูลข่าวสารสมุนไพรวัดป่าอรัญญิกาวาส.สมุนไพรสวนสมุนไพรวัดป่าอรัญญิกาวาส.มหาสารคาม: โครงการศูนย์ข้อมูลข่าวสารสมุนไพรวัดป่าอรัญญิกาวาส, 2537.

จันดี เข็มเฉลิม. รวบรวมตำรายาไทยที่ใช้ในพนมสารคาม. ฉะเชิงเทรา, 2525.

ชยันต์ วรรธนะภูติ.เวทีความคิด ศักยภาพและข้อจำกัดของหมอพื้นบ้านกับการดูแลผู้ป่วยเอดส์. สถาบันวิจัยสังคม มหาวิทยาลัยเชียงใหม่, 2537.

ชลอ อุทกภาชน์. คู่มือยาสมุนไพรและโรคประเทศร้อนและวิธีบำบัดรักษา. กรุงเทพมหานคร: หจก.โรงพิมพ์แพร่พิทยาอินเตอร์เนชั่นแนล, 2509.

ชลอ อุทกภาชน์. หลักการศึกษายาสมุนไพรและหลักการใช้ยาสมุนไพรรักษาโรคต่างๆ เปรียบเทียบกับการใช้ยาแผนปัจจุบันรักษาโรคเหล่านั้น. กรุงเทพมหานคร: หจก. อักษรธเนศวร, 2528.

เชษฐา พยากรณ์. สมุนไพรในชีวิตประจำวัน. กรุงเทพมหานคร: สำนักพิมพ์เชษฐา, 2525.

เชาวน์ กสิพันธุ์. ตำรายาโบราณ พระคัมภีร์กระษัยและยารักษาโรคกระษัย. กรุงเทพมหานคร: สมชายการพิมพ์, 2523.

เชาวน์ กสิพันธุ์. ตำราเภสัชศึกษา. กรุงเทพมหานคร: สมชายการพิมพ์, 2523.

เต็ม สมิตินันทน์, วีระชัย ณ นคร. พฤกษศาสตร์พื้นบ้าน. กรุงเทพมหานคร: หอพรรณไม้ กองบำรุง กรมป่าไม้, 2535.

เต็ม สมิตินันทน์. ชื่อพรรณไม้แห่งประเทศไทย. กรุงเทพมหานคร: หจก. ฟันนี่พับลิชชิ่ง, 2523.

ทวีทอง หงษ์วิวัฒน์. การวิจัยเพื่อพัฒนาการแพทย์แผนไทย:ทัศนะและการวิเคราะห์เชิงมหภาคเพื่อการใช้ประโยชน์. เอกสารประกอบการสัมมนาการแพทย์แผนไทย เรื่องคุณค่าและบทบาทของการแพทย์แผนไทยร่วมสมัย วันที่ 12-14 ตุลาคม 2535 ณ โรงแรมหาดทอง จังหวัดประจวบคีรีขันธ์.

ธารา อ่อนชมจันทร์และคณะ. รายงานการวิจัยทางเลือกในการรักษากระดูกหัก:กรณีศึกษาหมอกระดูก อำเภอพญาเม็งราย จังหวัดเชียงราย, 2537.

นันทวัน บุณยะประภัศร, บรรณาธิการ. ก้าวไปกับสมุนไพรเล่ม 2. กรุงเทพมหานคร: ธรรกมลการพิมพ์, 2530.

นันทวัน บุณยะประภัศร, บรรณาธิการ. ศัพท์แพทย์ไทย. กรุงเทพมหานคร: บริษัทประชาชน จำกัด, 2535.

ปรีชา อุยตระกูล และคณะ. บทบาทของหมอพื้นบ้านในสังคมชนบทอีสาน ศูนย์ข้อมูลท้องถิ่นเพื่อการพัฒนา. วิทยาลัยครูนครราชสีมา, 2531.

พร้อมจิต ศรลัมพ์. สมุนไพรออกฤทธิ์ต่อระบบทางเดินหายใจ. กรุงเทพมหานคร: คณะเภสัชศาสตร์ มหาวิทยาลัย มหิดล, 2530.

พระเทพวิมลโมลี. ตำรายากลางบ้านมีสรรพคุณชะงัด. กรุงเทพมหานคร: โรงพิมพ์มหามกุฏราชวิทยาลัย, 2524.

พิณ เกื้อกูล, พรทิพย์ ผลวิชา, พิพัฒน์ สุวิสิษฐ์, และคณะ บรรณาธิการ. พรรณไม้ในสวนป่าสิริกิติ์ ภาคกลาง(จังหวัด ราชบุรี). กรุงเทพมหานคร: ชุติมาการพิมพ์, 2535.

รุ่งระวี เต็มศิริฤกษ์กุล. พรรณไม้มีพิษ. กรุงเทพมหานคร: คณะเภสัชศาสตร์ มหาวิทยาลัยมหิดล, 2537.

รุ่งระวี เต็มศิริฤกษ์กุล. สมุนไพรรักษาโรคเรื้อรังบางชนิด. กรุงเทพมหานคร: คณะเภสัชศาสตร์ มหาวิทยาลัย มหิดล, 2537.

รุ่งรังษี วิบูลชัย. การดำรงอยู่ของการแพทย์พื้นบ้าน: กรณีศึกษาหมู่บ้านนาสีดา ตำบลข้าวปุ้น อำเภอกุดข้าวปุ้น จังหวัดอุบลราชธานี วิทยานิพนธ์ศิลปศาสตรมหาบัณฑิต(วัฒนธรรมศึกษา) มหาวิทยาลัยมหิดล, 2538.

รุจินาถ อรรถสิษฐ. การแพทย์พื้นบ้าน: มุมมองเพื่อการวิจัยและพัฒนา. เอกสารประกอบคำสอนวิชามานุษย วิทยาการแพทย์พื้นบ้าน สถาบันวิจัยภาษาและวัฒนธรรมเพื่อพัฒนาชนบท.

โรงเรียนแพทย์แผนโบราณ วัดพระเชตุพนวิมลมังคลารามราชวรวิหาร. ตำราประมวลหลักเภสัช. กรุงเทพ- มหานคร: ไพศาลศิลป์การพิมพ์, 2528.

ลือชัย ศรีเงินยวง, รุจินาถ อรรถสิษฐ. ศักยภาพหมอพื้นบ้านกับการสาธารณสุขมูลฐาน: ภาพรวม. กรุงเทพ- มหานคร: โรงพิมพ์องค์การสงเคราะห์ทหารผ่านศึก, 2535.

วงศ์สถิตย์ ฉั่วกุล. การสำรวจการใช้พันธุ์ไม้ของชนกลุ่มน้อยเผ่าซาไก ใน: สุรินทร์ ภู่ขจร,บรรณาธิการ. รายงาน เบื้องต้นการขุดค้นที่ถ้ำหมอเขียว จ.กระบี่, ถ้ำซาไก จ.ตรัง และการศึกษาชาติพันธุ์วิทยาทางโบราณคดี ชนกลุ่มน้อยเผ่าซาไก จ.ตรัง. กรุงเทพมหานคร: โครงการวิจัยวัฒนธรรมโหบินเนียนในประเทศไทย, 2534: 102-112.

วงศ์สถิตย์ ฉั่วกุล, พร้อมจิต ศรลัมพ์, รุ่งระวี เต็มศิริฤกษ์กุล. สมุนไพรพื้นบ้าน อำเภอกาบเชิง จังหวัดสุรินทร์. วารสารเภสัชศาสตร์ มหาวิทยาลัยมหิดล 2537; 21(3): 102-112.

วงศ์สถิตย์ ฉั่วกุล, ยุวดี วงษ์กระจ่าง, จันทร์เพ็ญ วิวัฒน์ และคณะ. การศึกษาการใช้สมุนไพรของชนกลุ่มน้อย เผ่ามลาบรี, ถิ่น, ขมุ และเย้า.ใน: สุรินทร์ ภู่ขจร,บรรณาธิการ. ผลการวิเคราะห์กลุ่มสังคมล่าสัตว์ เผ่า "ผีตองเหลือง" ในประเทศไทย. กรุงเทพมหานคร: บริษัททอมรินทร์พริ้นติ้งกรุ๊พ จำกัด, 2531: 118-144.

วงศ์สถิตย์ ฉั่วกุล, สมภพ ประธานธุรารักษ์, พร้อมจิต ศรลัมพ์ และคณะ. สมุนไพรพื้นบ้าน จังหวัดมหาสารคาม. วารสารสมุนไพร 2538; 1(2): 39-56.

สมพร ภูติยานันต์. การสำรวจการใช้สมุนไพรของแพทย์แผนโบราณ. กรุงเทพมหานคร: โรงพิมพ์พิฆเณศ, 2521.

สมาคมป่าไม้แห่งประเทศไทย. ไม้และของป่าบางชนิดในประเทศไทย. กรุงเทพมหานคร: หจก.นำอักษรการพิมพ์, 2527.

สมาคมโรงเรียนแพทย์แผนโบราณ วัดพระเชตุพนฯ. ประมวลสรรพคุณยาไทย ภาค 1. กรุงเทพมหานคร, 2521.

สมาคมโรงเรียนแพทย์แผนโบราณ วัดพระเชตุพนฯ. ประมวลสรรพคุณยาไทย ภาค 2. กรุงเทพมหานคร, 2521.

สมาคมโรงเรียนแพทย์แผนโบราณ วัดพระเชตุพนฯ. ประมวลสรรพคุณยาไทย ภาค 3. กรุงเทพมหานคร, 2521.

สอาด บุญเกิด, จเร สดากร, ทิพย์พรรณ สดากร. ชื่อพรรณไม้ในเมืองไทย. กรุงเทพมหานคร: พ.จิระการพิมพ์, 2525.

สายสนม กิตติขจร. ตำราสรรพคุณสมุนไพรยาไทยแผนโบราณ. กรุงเทพมหานคร: โรงพิมพ์อักษรไทย, 2526.

สำนักงานคณะกรรมการการสาธารณสุขมูลฐาน. การสำรวจการยอมรับการแพทย์พื้นบ้านของบุคลากร
    สาธารณสุข. สำนักงานปลัดกระทรวงสาธารณสุข, 2532.

สุนทรี สิงหบุตรา. สรรพคุณสมุนไพร 200 ชนิด. กรุงเทพมหานคร: บริษัทคุณ39 จำกัด, 2535.

สุวิไล เปรมศรีรัตน์. วิธีป้องกันและรักษาโรคแบบพื้นบ้านชาวขมุและสนทนาสาธารณสุข การแพทย์ไทย-ขมุ.
    กรุงเทพมหานคร: โรงพิมพ์มหาจุฬาลงกรณราชวิทยาลัย, 2533.

เสงี่ยม พงษ์บุญรอด. ไม้เทศเมืองไทย. กรุงเทพมหานคร: เกษมบรรณกิจ, 2522.

เสาวภา พรสิริพงษ์. การแพทย์พื้นบ้านกับสมุนไพร. การสัมมนาวิชาการเรื่องการแพทย์แผนไทยกับสังคมไทย
    โครงการจัดงานทศวรรษการแพทย์แผนไทย วันที่ 10-13 มีนาคม 2538 ณ ศูนย์การประชุมแห่งชาติสิริกิติ์.

อวย เกตุสิงห์. การแพทย์ไทยกับการแพทย์ตะวันตก. วารสารสังคมศาสตร์การแพทย์. 2521: 1(2).

อิริค ไซเดนฟาเดน. ชาติวงศ์วิทยา: ว่าด้วยชนชาติเผ่าต่างๆในประเทศ. กรุงเทพมหานคร: อักษรเจริญทัศน์, 2509.

Anderson EF. Ethnobotany of hill tribes of Northern Thailand.I. Medicinal plants of Akha. Econ Bot 1986;
    40(1): 38-53.

Anderson EF. Ethnobotany of hill tribes of Northern Thailand. II. Lahu medicinal plants. Econ Bot 1986; 40(4):
    442-450.

Backer CA and Bakhuizen Van Den Brink Jr RC. Flora of Java, vol I. Groningen : NVP Noordhoff, 1963.

Backer CA and Bakhuizen Van Den Brink Jr RC. Ibid vol II. Groningen : NVP Noordhoff, 1963.

Backer CA and Bakhuizen Van Den Brink Jr RC. Ibid vol III. Groningen : NVP Noordhoff, 1968.

Bjornland T, Schumacher T. Compilation of ethnobotanical data In: Brun V, Schumacher T. Traditional herbal
    medicine in Northern Thailand. California: University of California Press,1987: 241-302.

Bye RA Jr. Botanical perspectives of ethnobotany of the Greater Southwest. Econ Bot 1985; 39(4): 375-386.

Chuakul W. *Jasminum rottlerianum* Wall. ex DC., the new recorded species of Thailand. Mahidol Univ J
    Pharm Sci 1994; 21(1): 32-34.

Collins DJ, Culvenor CCJ, Lamberton JA, et al. Plants for medicines. Melbourne: Brown Prior Anderson
    Pty ltd, 1990.

Cox PA. Ethnopharmacology and the search of new drugs In: Ciba Foundation Symposium 154. Bioactive
    compound from plants. New York: John Wiley & Sons, 1990: 40.

Dassanayake MD, Forsberg FR, eds. A revised handbook to the Flora of Ceylon. Vol.1. New Delhi: Oxford
    & IBH Publishing Co., 1980.

Dassanayake MD, Forsberg FR, eds. Ibid Vol.3. New Delhi: Oxford & IBH Publishing Co., 1981.

Dassanayake MD, Forsberg FR, eds. Ibid Vol.4. New Delhi: Oxford & IBH Publishing Co., 1983.

Dassanayake MD, Forsberg FR, eds. Ibid Vol.5. New Delhi: Oxford & IBH Publishing Co., 1985.

Dassanayake MD, Forsberg FR, eds. Ibid Vol.6. New Delhi: Oxford & IBH Publishing Co., 1987.

Dassanayake MD, Forsberg FR, eds. Ibid Vol.7. New Delhi: Oxford & IBH Publishing Co., 1991.

Dunn FL. Traditional Asian medicine and cosmopolitan medicine as adaptive systems In: C. Leslie, ed. Asian medicalsystems: a comparative study. Los Angeles: University of California Press, 1977.

Farnsworth NR. The role of ethnopharmacology in drug development In: Ciba Foundation Symposium 154. Bioactive compound from plants. New York: John Wiley & Sons, 1990: 2.

Ford RI. Anthropological perspective of ethnobotany in the Greater Southwest. Econ Bot 1985; 39(4): 400-.415.

Foster GM, Anderson BG. Medical Antropology. New York: John Wiley & Son, 1978.

Holttum RE. A Revised Flora of Malaya. Vol. 1. Singapore: The Government Printing Office, 1964.

Iwu MM. Handbook of African medicinal plants. Boca Raton: CRC Press, 1993.

Larsen K, Smitinand T, eds. Flora of Thailand. Vol.2 Part 2. Bangkok: The ASRCT Press, 1972.

Larsen K, Smitinand T, eds. Ibid Vol.2 Part 3. Bangkok: The ASRCT Press, 1975.

Larsen K, Smitinand T, eds. Ibid Vol.2 Part 4. Bangkok: The TISTR Press, 1981.

Larsen K, Smitinand T, eds. Ibid Vol.3 Part 1. Bangkok: The TISTR Press, 1990.

Larsen K, Smitinand T, eds. Ibid Vol.4 Part 2. Bangkok: The TISTR Press, 1985.

Larsen K, Smitinand T, eds. Ibid Vol.5 Part 1. Bangkok: The Chutima Press, 1987.

Larsen K, Smitinand T, eds. Ibid Vol.5 Part 2. Bangkok: The Chutima Press, 1990.

Larsen K, Smitinand T, eds. Ibid Vol.5 Part 3. Bangkok: The Chutima Press, 1991.

Larsen K, Smitinand T, eds. Ibid vol.5 Part 4. Bangkok: The Chutima Press, 1992.

Larsen K, Smitinand T, eds. Ibid Vol.6 Part 1. Bangkok: The Rumthai Press, 1993.

Lipp FJ. Methods for ethnopharmacological field work. J Ethnopharmacol 1989; 25: 139-150.

Moerman DE. Symbols and selectivity: A statistical analysis of native American medical ethnobotany. J of Ethnopharmacol 1979; 1: 111-119.

Peigen X. Recent developments on medicinal plants in China. J Ethnopharmacol 1983; 7: 95-119.

Perry, LM. Medicinal plants of East and Southeast Asia. London: The MIT Press, 1980.

Phengklai C, Niyomdham C. Flora in peat swamp areas of Narathiwat. Bangkok: S. Sombun Press, 1991.

Radanachaless T, Maxwell JF. Weeds of soybean fields in Thailand. Chaing Mai: Multiple Cropping Center, 1994.

Rafatullah S, Tariq M, Al-Yahya MA, Mossa JS, Ageel AM. Evaluation of Turmeric (Curcuma longa) for gastric and duodenal antiulcer activity in rats. J Ethnopharmacol 1990; 29: 25-34.

Schultes RV. The role of the ethnobotanist in the search for new medicinal plants. Lloydia 1962; 25(4): 257-266.

Tang. W, Eisenbrand G. Chinese drugs of plants origin. Berlin Heidelberg : Springer Verlag, 1992.

The Council of The Royal Pharmaceutical Society of Great Britain. Martindale : The Extra Pharmacopoeia. 29th ed. London: The Pharmaceutical Press, 1989.

Trease GE, Evans WC. Pharmacognosy. 12th ed. Oxford: Alden press, 1985.

Tyler VE, Brady LR, Robbers JE. Pharmacognosy. 8 th ed. Philadelphia: Lea & Febiger, 1981.

Tyler VE. Herbs of choice : The therapeutic use of phytomedicinals. New York: The Haworth Press, Inc., 1994.

Van Steenis CGGJ, ed. Flora Malesiana. Volume 4. New Delhi: Oxford & IBH Publishing Co., 1950.

Van Steenis CGGJ, ed. Ibid Volume 5. New Delhi: Oxford & IBH Publishing Co., 1950.

Van Steenis CGGJ, ed. Ibid Volume 6 Part 1. New Delhi: Oxford & IBH Publishing Co., 1950.

Van Steenis CGGJ, ed. Ibid Volume 6 Part 3. New Delhi: Oxford & IBH Publishing Co., 1950.

Van Steenis CGGJ, ed. Ibid Volume 8 Part 2. New Delhi: Oxford & IBH Publishing Co., 1950.

Van Steenis CGGJ, ed. Ibid Volume 8 Part 3. New Delhi: Oxford & IBH Publishing Co., 1978.

Xolocotzi EH. Experiences leading to a greater emphasis on man in ethnobotanical studies. Econ Bot 1987;
    41(1): 6-11.

INDEX OF ENGLISH NAMES

APPENDIX
A list of medicinal plants grouping in families

ACANTHACEAE

Clinacanthus siamensis Brem.

Graptophyllum pictum Griff.

AGAVACEAE

Cordyline fruticosa Goppert

ALLIACEAE

Allium tuberosum Roxb.

ALTINGIACEAE

Altingia siamensis Craib

AMARANTHACEAE

Cyathula prostrata (Linn.) Bl.

ANNONACEAE

Annona reticulata Linn.

Cananga latifolia Finet et Gagnep.

Cyathostemma micranthum (A. DC.) J. Sincl.

Polyalthia cerasoides (Roxb.) Benth. ex Bedd.

Polyalthia debilis (Pierre) Finet et Gagnep.

Polyalthia evecta (Pierre) Finet et Gagnep.

Polyalthia suberosa Thw.

Uvaria rufa Bl.

APOCYNACEAE

Aganonerion polymorphum Pierre ex Spire

Aganosma marginata G. Don

Carissa carandas Linn.

Carissa cochinchinensis Pierre

Cerbera odollam Gaertn.

Strophanthus caudatus (Linn.) Kurz

(*S. scandens* (Lour.) Roem. et Schult.)

Strophanthus gratus Franch.

ARACEAE

Lasia spinosa Thw.

Typhonium trilobatum Schott

ARALIACEAE

Trevesia palmata Vis.

ARISTOLOCHIACEAE

Aristolochia pothieri Pierre ex Lec.

Thottea parviflora Ridl.

T. tomentosa (Bl.) Ding Hou

ASCLEPIADACEAE

Asclepias curassavica Linn.

Calotropis gigantea (Linn.) R. Br. ex Ait.

Dischidia major (Vahl) Merr.

(*D. rafflesiana* Wall.)

Oxystelma secamone (Linn.) Karst.

(*O. esculentum* (Linn.) R. Br.)

Raphistemma hooperianum Decne.

Sarcostemma brunonianum Wight et Arn.

ASPHODELACEAE

Asparagus racemosus Willd.

ASTERACEAE

Blumea balsamifera (Linn.) DC.

Chromolaena odorata (Linn.) King et Robins.

    (*Eupatorium odoratum* Linn.)

Eclipta prostrata Linn.

AVERRHOACEAE

Averrhoa bilimbi Linn.

BALANOPHORACEAE

Aeginetia indica Roxb.

BARRINGTONIACEAE

Careya sphaerica Roxb.

CAMPANULACEAE

Laurentia longiflora Peterm.

CAPPARACEAE

Capparis micracantha DC.

Crateva adansonii DC.

    ssp. *trifoliata* (Roxb.) Jacobs

Crateva religiosa Ham.

Maerua siamensis Kurz

CASSYTHACEAE

Cassytha filiformis Linn.

CELASTRACEAE

Celastrus paniculatus Willd.

Maytenus marcanii Ding Hou

Salacia chinensis Linn.

CLEOMACEAE

Cleome viscosa Linn.

CLUSIACEAE

Calophyllum inophyllum Linn.

Cratoxylum formosum (Jack) Dyer

Garcinia schomburgkiana Pierre

COMBRETACEAE

Terminalia citrina Roxb. ex Flem.

CONNARACEAE

Connarus semidecandrus Jack

Ellipanthus tomentosus Kurz ssp. *tomentosus*

    var. *tomentosus*

CONVOLVULACEAE

Argyreia nervosa (Burm. f.) Boj.

Ipomoea quamoclit Linn.

Neuropeltis racemosa Wall.

COSTACEAE

Costus speciosus (Koen.) J.E. Smith

CUCURBITACEAE

Cucurbita moschata Decne.

Momordica cochinchinensis (Lour.) Spreng.

Mukia maderaspatana Roem.

Solena heterophylla Lour.

CYCADACEAE

Cycas pectinata Griff.

DICKSONIACEAE

Cibotium barometz J. Smith

DIOSCOREACEAE

Dioscorea hispida Dennst.

DIPTEROCARPACEAE

Dipterocarpus obtusifolius Teijsm. ex Miq.

Hopea odorata Roxb.

EBENACEAE

Diospyros decandra Lour.

Diospyros rhodocalyx Kurz

EQUISETACEAE

Equisetum debile Roxb.

ERYTHROXYLACEAE

Erythroxylum cuneatum (Miq.) Kurz

EUPHORBIACEAE

Aporusa villosa (Lindl.) Baill.

Baliospermum montanum (Willd.) Muell. Arg.

Cladogynos orientalis Zipp. ex Span.

Croton crassifolius Giesel

Croton oblongifolius Roxb.

Euphorbia hirta Linn.

Glochidion lanceolarium (Roxb.) Voigt

Mallotus philippensis (Lam.) Muell. Arg.

Phyllanthus acidus (Linn.) Skeels

Strophioblanchia glandulosa Pax

    var. *pandurifolia* Airy Shaw

Trigonostemon reidioides (Kurz) Craib

FABACEAE

Afgekia sericea Craib

Afzelia xylocarpa (Kurz) Craib

Bauhinia pulla Craib

Caesalpinia mimosoides Lamk.

Cajanus cajan Millsp.

Cassia bakeriana Craib

Cassia sophera Linn.

Dendrolobium lanceolatum (Dunn.) Schindl.

Dendrolobium thorelii (Gagnep.) Schindl.

Desmodium gangeticum DC.

Hegnera obcordata (Miq.) Schindl.

Lespedeza parviflora Kurz

Mimosa invisa Mart. ex Colla

Mimosa pudica Linn. var. *hispida* Bren.

Parkia speciosa Hassk.

Peltophorum dasyrachis (Miq.) Kurz ex Baker

Phyllodium elegans (Lour.) Desv.

Phyllodium longipes (Craib) Schindl.

Phyllodium pulchellum (Linn.) Desv.

Pithecellobium tenue Craib

Psophocarpus tetragonolobus DC.

Pterocarpus macrocarpus Kurz

Saraca declinata (Jack) Miq.

Senna timoriensis (DC.) H.S. Irwin et

    R.C. Barneby

  (*Cassia timoriensis* DC.)

Sindora siamensis Teijsm. ex Miq. var. *siamensis*

Tadehagi godefroyanum (O. Ktze.) Ohashi

Xylia xylocarpa (Roxb.) Taub.

    var. *kerrii* (Craib et Hutch.) Nielsen

FLACOURTIACEAE

Casearia grewiaefolia Vent.

LAMIACEAE

Hyptis suaveolens (Linn.) Poit.

Leonurus sibiricus Linn.

LAURACEAE

Litsea glutinosa (Lour.) C. B. Robinson

LEEACEAE

Leea indica Merr.

Leea macrophylla Roxb. ex Hornem.

LILIACEAE

Dianella ensifolia (Linn.) DC.

LORANTHACEAE

Dendrophthoe pentandra Miq.

MAGNOLIACEAE

Michelia alba DC.

MALPIGHIACEAE

Hiptage benghalensis Kurz

MALVACEAE

Abelmoschus moschatus Medic.

Abutilon indicum Sweet

Abutilon polyandrum G. Don

Decaschistia parviflora Kurz

Hibiscus mutabilis Linn.

MARANTACEAE

Donax grandis Ridl.

MEMECYLACEAE

Memecylon edule Roxb.

MENISPERMACEAE

Stephania venosa (Bl.) Spreng.

MORACEAE

Ficus benjamina Linn.

Ficus foveolata Wall.

Ficus racemosa Linn.

Morus alba Linn.

Streblus asper Lour.

MYRTACEAE

Melaleuca leucadendra Linn. var. *minor* Duthie

Rhodamnia dumetorum DC.) Merr. et Perry

NYCTAGINACEAE

Pisonia aculeata Linn.

OCHNACEAE

Gomphia serrata (Gaertn.) Kanis

OLACACEAE

Olax scandens Roxb.

OLEACEAE

Jasminum rottlerianum Wall. ex DC.

Linociera macrophylla Wall.

(*L. ramiflora* Wall. ex G. Don)

OPILIACEAE

Lepionurus sylvestris Bl.

ORCHIDACEAE

Calanthe cardioglossa Schltr.

Cymbidium aloifolium (Linn.) Swartz

(*O. simulans* Rolfe)

Dendrobium draconis Reichb. f.

Dendrobium trigonopus Reichb. f.

Grammatophyllum speciosum Bl.

Vanilla planifolia Andr.

OXALIDACEAE

Biophytum sensitivum DC.

PERIPLOCACEAE

Cryptolepis buchanani Roem. et Schult.

POLYGONACEAE

Muehlenbeckia platyclada (F.v.Muell.) Meissn.

ROSACEAE

Parinari anamensis Hance

RUBIACEAE

Catunaregam tomentosa (Bl. ex DC.) Tirveng.

(*Randia dasycarpa* Bakh. f.)

Cinchona succirubra Par.

Coffea canephora Pierre ex Froehner

Dioecercis erythroclada (Kurz) Tirveng.

(*Gardenia erythroclada* Kurz)

Gardenia sootepensis Hutch.

Geophila herbacea (Linn.) O. Ktze.

Hydnophytum formicarum Jack

Mitragyna speciosa Korth.

Morinda elliptica Rindl.

Oxyceros horridus Lour.

(*Randia siamensis* Craib)

Rothmannia wittii (Craib) Bremek.

(*Randia wittii* Craib)

RUTACEAE

Clausena harmandiana Pierre ex Guill.

Feronia limonia Swing.

Glycosmis pentaphylla Corr.

Naringi crenulata (Roxb.) Nicolson

(*Hesperethusa crenulata* (Roxb.) Roem.)

Toddalia asiatica (Linn.) Lamk.

SAPINDACEAE

Arfeuillea arborescens Pierre

Cardiospermum halicacabum Linn.

Lepisanthes rubiginosa (Roxb.) Leenh.

Lepisanthes senegalensis (Poir.) Leenh.

SCROPHULARIACEAE

Lindenbergia philippensis Benth.

SIMAROUBACEAE

Picrasma javanica Bl.

SMILACACEAE

Smilax siamensis T. Koyama

SOLANACEAE

Nicotiana tabacum Linn.

Physalis minima Linn.

Solanum stramonifolium Jacq.

STEMONACEAE

Stemona phyllantha Gagnep.

STERCULIACEAE

Helicteres angustifolia Linn.

Theobroma cacao Linn.

STRYCHNACEAE

Strychnos lucida R. Br.

Strychnos nux-blanda A. W. Hill

SYMPHOREMACEAE

Sphenodesme involucrata (Presl) Robinson

TACCACEAE

Tacca integrifolia Ker-Gawl.

THEACEAE

Schima wallichii (DC.) Korth.

THYMELAEACEAE

Enkleia siamensis Nervling

Linostoma decandrum (Roxb.) Wall. ex Meissn.

TILIACEAE

Colona auriculata (Desv.) Craib

ULMACEAE

Trema orientalis (Linn.) Bl.

VERBENACEAE

Clerodendrum infortunatum Gaertn.

Clerodendrum paniculatum Linn.

Clerodendrum serratum (Linn.) Moon
var. *serratum* Schau.

Stachytarpheta indica Vahl

Tectona grandis Linn.

Vitex glabrata R. Br.

Volkameria fragrans Vent.

Zingiber ottensii Val.

VITACEAE

Cayratia trifolia (Linn.) Domin.

ZINGIBERACEAE

Hedychium coronarium Roem.

ZYGOPHYLLACEAE

Tribulus terrestris Linn.

AUTHORS

Thai Folk Medicine

Arthorn Riewpaiboon

Aim-on Somanabandhu

Thai Medicinal Plants

Wongsatit Chuakul

Promjit Saralamp

Wichit Paonil

Rungravi Temsiririrkkul

Arthorn Riewpaiboon

Sompop Prathanturarug

Thai-Style Herbal Steam Bath

Payow Muenwongyati

Aim-on Somanabandhu